A HOLE IN THE

A Hole in the Head

More Tales in the History of Neuroscience

———

Charles G. Gross

The MIT Press

Cambridge, Massachusetts

London, England

First MIT Press paperback edition, 2012

© 2009 Massachusetts Institute of Technology

MIT Press books may be purchased at special quantity discounts for business or sales promotional use. For information, please e-mail special_sales@mitpress.mit.edu or write to Special Sales Department, The MIT Press, 55 Hayward Street, Cambridge, MA 02142.

This book was set in Bembo on 3B2 by Asco Typesetters, Hong Kong. Printed and bound in the United States of America.

Library of Congress Cataloging-in-Publication Data

Gross, Charles G.
A hole in the head : more tales in the history of neuroscience / Charles G. Gross.
    p.   cm.
Includes bibliographical references and index.
ISBN 978-0-262-01338-3 (hardcover : alk. paper)—978-0-262-51733-1 (pb.)

1. Neurosciences—History. 2. Neurosciences and the arts. 3. Neuroscientists. I. Title.
RC338.G76   2009
616.80092′2—dc22   2009009928

10   9   8   7   6   5   4   3

To Joyce Carol Oates

CONTENTS

My interest in the history of neuroscience began when I was an undergraduate and extended for more than 50 years throughout my career as an experimental neuroscientist specializing in vision and the functions of the cerebral cortex. This is my second collection of history of neuroscience papers; the first, published in 1998, was entitled *Brain, Vision, Memory: Tales in the History of Neuroscience*.

This collection differs from the previous one in including postscripts that endeavor to bring the papers up date, even those that deal with events over a thousand years ago. For example both the study of trephining, cutting holes in the skull, and of the Greek physician and scientist Galen of Pergamon remain active and often controversial fields of research with continuing interesting new developments.

Part I deals with early developments in neuroscience that are still relevant today. The articles in part II deal with the interactions of art and science. Part III tells the stories of neuroscientists whose ideas were so advanced as to be ignored or even inconceivable to their contemporaries.

In general, for the chapters previously published I have made only a few minor corrections and changes other than to eliminate the abstracts, add cross-references, make the citations consistent, and provide a consolidated reference list.

Chapter 6 was coauthored with Marc Bornstein and Chapter 9 with Michael Colombo and his father Arnaldo Colombo. Chapter 4 arose, in part, from a collaborative effort with Charlotte Taylor.

Many friends and colleagues made suggestions or helped with the original articles or postscripts including T. Albright, S. Alisharan, M. Benti-voglio, G. Berman, B. Campbell, D. Cooke, J. Cooper, J. E. Cottrell, R. Desimone, R. Galambos, S. Gandhi, A. Ghanzanafer, E. Gould, M. Graziano, A. Grinnell, D. A. Gross, M. Hauser, F. L. Holmes, X. T. Hu, E. Isaac, G. Krauthammer, S. Landes, M. Mugan, J. C. Oates, R. Payne, G. E. Peierls, R. E. Peierls, A. Repp, H. R. Rodman, L. Seacord, T. Sejnowski, J. Simmons, M. Sommer, H. Terrace, S. Waxman, G. Winer, and R. M. Young.

Linda Chamberlin of the Princeton University library was tireless in getting books and articles. Shalani Alisharan copyedited magnificently. Tyler Clark played a major role in every aspect of the preparation of this book. Production of the original articles was helped by my grants from the National Institutes of Health EY 11347-1-34.

I

EARLY NEUROSCIENCE AND ITS REVERBERATIONS TODAY

Modern neuroscience is often said to have begun with the work of the great neuroanatomist Ramón y Cajal (1852–1934) and that of the founder of modern neurophysiology E. D. (Lord) Adrian (1889–1977). Yet the study of brain function has much deeper historical roots. For example, the author of the Edwin Smith Surgical Papyrus, written about 1000 BCE (but derived from an older treatise from about 3000 BCE), knew that different types of head injury produce different symptoms. At the very beginning of formal science in the sixth century BCE, several of the pre-Socratic natural philosophers (particularly the ones who were also physicians) were aware of the hegemony of the brain in sensation, movement, and mentation.

The four chapters in this part of the book span this long history. The first deals with trepanation, or cutting holes in the skull. It was practiced among the Neolithic Peruvian Indians and, through subsequent centuries, all over the world; it is advocated even today on the Internet for "enhancing" consciousness.

The second chapter deals with an experimental demonstration in classical times that was aimed at proving the role of the brain as opposed to the heart in cognition and emotion. The controversy between head and heart as the seat of mind stretches from the time of preliterate cultures well into the Renaissance.

The third discusses another question that dates to ancient Greece, namely whether vision involves something going into the eye or coming from it—a question that seems still unresolved among contemporary undergraduates. This question bears on both the "evil eye" and the feeling of being stared at.

The final chapter in this part concerns a more modern problem, one that has extended from the nineteenth century until today: the role of the cerebral cortex in producing movement of the body.

A Hole in the Head: A History of Trepanation

The oldest known surgical procedure is trepanation or trephination, the removal of a piece of bone from the skull. It was practiced from the late Paleolithic period and in virtually every part of the world. It is still used in both Western and non-Western medicine. We consider the methods and motives of trephining in different times and cultures.

A Peruvian Skull

In 1865, in the ancient Inca city of Cuzco, Ephraim George Squier, explorer, archeologist, ethnologist and U.S. *charge d'affaires* in Central America, received an unusual gift from his hostess, Señora Zentino, a woman known as the finest collector of art and antiquities in Peru. The gift was a skull from a vast nearby Inca burial ground. What was unusual about the skull was that a hole slightly larger than a half-inch square had been cut out of it (see figure 1.1). Squier's judgment was that the skull hole was not an injury but was the result of a deliberate surgical operation known as trepanning and furthermore, that the individual had survived the surgery.[1]

When the skull was presented to a meeting of the New York Academy of Medicine, the audience refused to believe that anyone could have

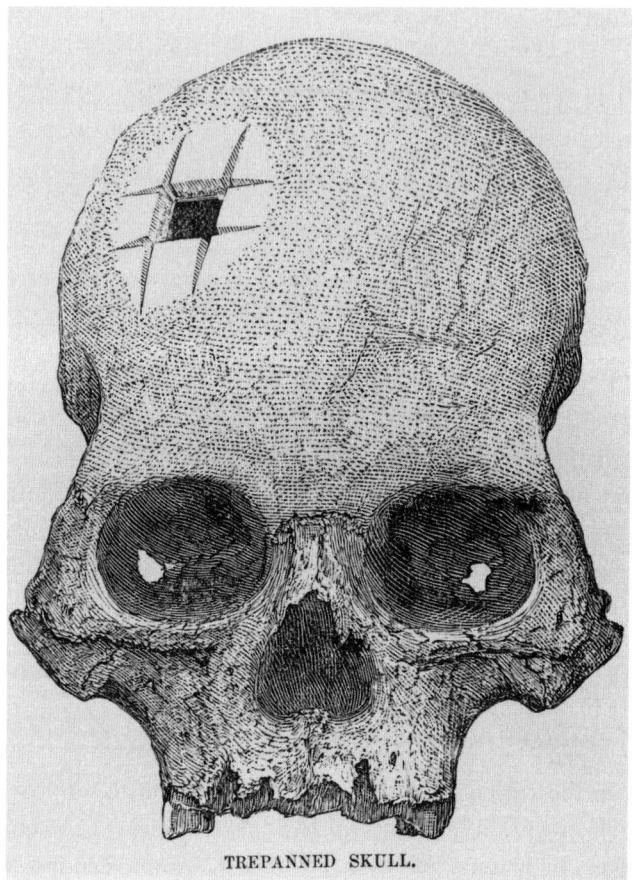

TREPANNED SKULL.

Figure 1.1
The trephined Inca skull given to Squier, showed to Broca, and now residing in the American Museum of Natural History (Squier, 1877).

survived a trephining operation carried out by a Peruvian Indian.[2] Aside from the racism characteristic of the time, the skepticism was fueled by the fact that in the very best hospitals of the day, the survival rate from trephining (and many other operations) rarely reached 10%, and thus the operation was viewed as one of the most perilous surgical procedures.[3] The main reason for the low survival rate was the deadly infections then rampant in hospitals. Another was that the operation was only attempted in very severe cases of head injury.

Squier then brought his Peruvian skull to Europe's leading authority on the human skull, Paul Broca, professor of external pathology and of clinical surgery at the University of Paris and founder of the first anthropological society. Today, of course, Broca is best known for his localization of speech in the third frontal convolution, "Broca's area," the first example of cerebral localization of a psychological function, but at this time his fame seems to have been primarily for his craniometric and anthropological studies.[4]

BROCA AND MORE SKULLS

After examining the skull and consulting some of his surgical colleagues, Broca was certain that the hole in the skull was due to trephination and the patient had survived for a while. But when, in 1876, Broca reported these conclusions to the Anthropological Society of Paris, the audience, as in the United States, was dubious that Indians could have carried out this difficult surgery successfully.[5]

Seven years later a discovery was made in central France that confirmed Broca's interpretation of Squier's skull, or at least, demonstrated that "primitives," indeed Neolithic ones, could trephine successfully. A number of skulls in a Neolithic gravesite were found with roundish holes 2 or 3 inches wide. The skulls had scalloped edges as if they had been scraped with a sharp stone. Even more remarkable, discs of skull of the same size as the holes were found in these sites. Some of the discs had small holes bored in them, perhaps for stringing as amulets. Although a few of the discs had been chiseled out after death, in most cases it was clear from the scar formation at the wound's edge that the interval between surgery and death must have been years. Trephined skulls were found of both genders and of all ages. Virtually none of the skull holes in this sample were accidental, pathological, or traumatic. Furthermore very few of the skulls showed any sign of depressed fractures, a common indication for trephining in modern times.[6]

These findings finally established that Neolithic man could carry out survival trephination but left unresolved the motivation for this operation. At first, Broca thought that the practice must have been some kind of religious ritual, but later he concluded that, at least in some cases, it must have had therapeutic significance. Broca actually wrote more papers on prehistoric trephination and its possible motivation than he did on the cortical localization of language.[7] Since Broca's time thousands of trephined skulls have been found and almost as many papers written about them. They have been discovered in widespread locations in every part of the world in sites dating from the late Paleolithic to this century. The usual estimates for survival of different samples of trephined skulls ranges from 50% to 90% with most estimates on the higher side.[8]

METHODS OF TREPHINING

Across time and space five main methods of trephination were used.[9] The first was rectangular intersecting cuts as in Squier's skull (figures 1.1 and 1.2). These were first made with obsidian, flint, or other hard stone knives and later with metal ones. Peruvian burial sites often contain a curved metal knife called a tumi, which would seem to be well suited for the job. (The tumi has been adopted by the Peruvian Academy of Surgery as its emblem.) In addition to Peru, skulls trephined with this procedure have been found in France, Israel, and Africa.

The second method was scraping with a flint as in skulls found in France and studied by Broca. Broca demonstrated that he could reproduce

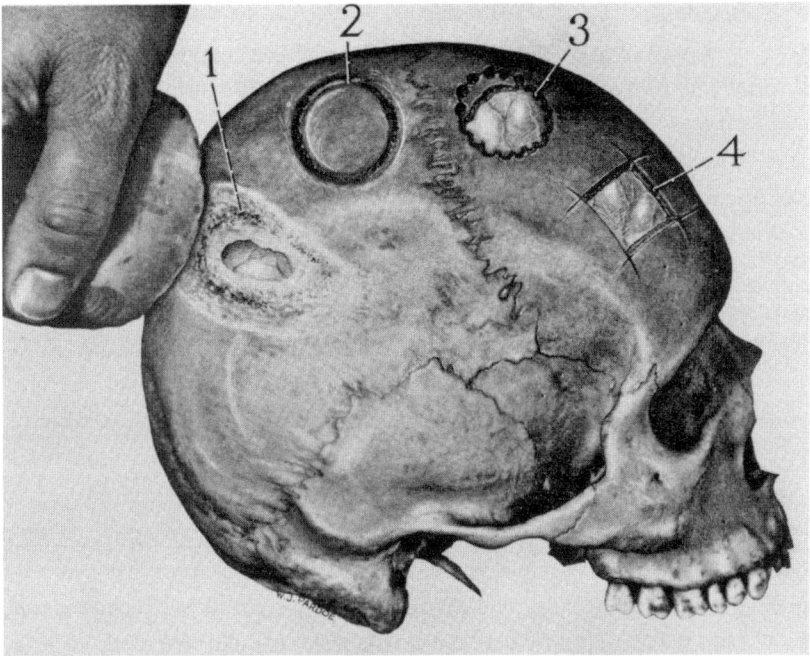

Figure 1.2
Different methods of trephining: (1) scraping; (2) grooving; (3) boring and cutting; (4) rectangular intersecting cuts (Lisowski, 1967).

these openings by scraping with a piece of glass, although a very thick adult skull took him 50 minutes "counting the periods of rest due to fatigue of the hand."[10] This was a particularly common method and persisted into the Renaissance in Italy.

The third method was cutting a circular groove and then lifting off the disc of bone. This is another common and widespread method and was still in use, at least until recently, in Kenya.

The fourth method, the use of a circular trephine or crown saw, may have developed out of the third. The trephine is a hollow cylinder with a toothed lower edge. Its use was described in detail by Hippocrates.[11] By the time of Celsus, a first-century Roman medical writer, it had a retractable central pin and a transverse handle. It looked almost identical to modern trephines including the one I used as a graduate student on monkeys.[12] (See figure 1.3.)

The fifth method was to drill a circle of closely spaced holes and then cut or chisel the bone between the holes. A bow may have been used for drilling or the drill simply rotated by hand. This method was recommended by Celsus, was adopted by the Arabs, and became a standard method in the Middle Ages. It is also reported to have been used in Peru and, until recently, in North Africa. It is essentially the same as the modern method for turning a large osteoplastic flap in which a Gigli saw (a sharp-edged wire) is used to saw between a set of small trephined or drilled holes.[13] (I used this method as a graduate student, too.)

"TREPAN" VERSUS "TREPHINE"

The relationship between the terms *trepan* and *trephine* is a curious one. The terms are now synonyms but have different origins and once had different meanings. In Hippocrates' time the terms *terebra* and *trepanon* (from the Greek *trupanon*, a borer) were used for the instrument that is very similar to the modern trephine. In the sixteenth century Fabricius ab Aquapendente invented a triangular instrument for boring holes in the skull. (He was

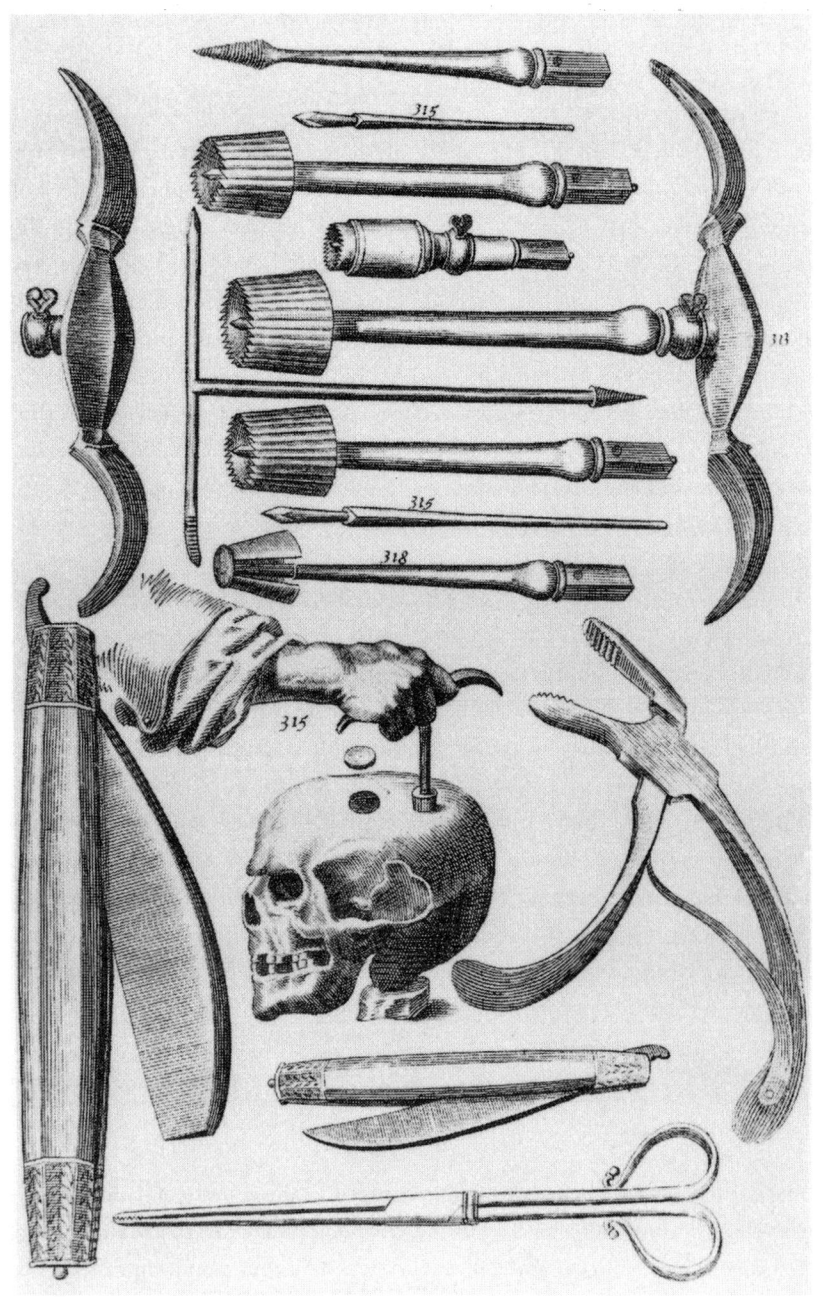

Figure 1.3
A seventeenth-century naval surgeon's trephination kit (Woodall, 1639). The trephines
are very similar to both ancient Roman and modern ones (Wilkins, 1997).

Harvey's teacher and the discoverer of venous valves.) It had three arms with different-shaped points. Each of the ends could be applied to the skull using the other two as handles. He called it a "tre fines" from the Latin for three ends, which became *trafine* and then *trephine*, and by 1656 it was used as a synonym for *trepan*, as a term for the older instrument. In another version of the etymology, a quite different triangular instrument for boring a hole in the skull was invented in 1639 by John Woodall, a London surgeon, who also called his instrument a tres fines, which became *trefina* and then *trephine* and, eventually, a synonym for *trepan*. More generally, in Renaissance times and later, trephination was a popular operation and a great variety of instruments for carrying it out were invented.[14]

WHY TREPHINE?

Why did so many cultures in different periods cut or drill holes in the skull? Since most trephined skulls come from vanished nonliterate cultures, the problem of reconstructing the motivations for trephining in these cultures is a difficult one. However, there is information about trephining in Western medicine from the fifth century BCE onward as well as about trephining in recent and contemporary non-Western medical systems. Both of these sources may throw light on the reasons for the practice in earlier times. In the following sections we consider trephination in Hippocratic medicine, in ancient Chinese medicine, in European medicine from the Renaissance onward, in contemporary non-Western medicine, and on the Internet today.

GREEK MEDICINE

The earliest detailed account of trephining is in the Hippocratic corpus, the first large body of Western scientific or medical writing that has survived. Although there is no question that there was a famous physician called Hippocrates in the fifth century BCE, it is not clear which of the Hippocratic works were written by him. The most extensive discussion of head injuries and the use of trephining in their treatment is in the Hippocratic work *On Wounds in the Head*.[15]

This treatise describes five types of head wounds. Interestingly, however, the only type for which trephination is not advocated is in cases of depressed fractures. Even when there is not much sign of bruising, drilling a hole in the head is recommended. The trephining instrument was very similar to the modern trephine, except that it was turned between the hands or by a bow and string rather than by using a crosspiece. The Hippocratic writer stressed the importance of proceeding slowly and carefully in order to avoid injuring the [dural] membrane. Additional advice was to "plunge [the trephine] into cold water to avoid heating the bone . . . often examine the circular track of the saw with the probe. . . . [and] aim at to and fro movements."[16] Trephining over a suture was to be studiously avoided.

Apparently the Hippocratic doctors expected bleeding from a head wound and the reason for drilling the hole in the skull was to allow the blood to escape ("let blood by perforating with a small trepan, keeping a look out [for the dura] at short intervals"). Since they presumably had no notion of intracerebral pressure, why did they want the blood to run out? Although the reasons for trephining are not discussed in *On Wounds in the Head*, they seem clear from other Hippocratic treatises such as *On Wounds* and *On Diseases*. The Hippocratic doctors believed that stagnant blood (like stagnant water) was bad. It could decay and turn into pus. Thus, the reason for trephining, or at least one reason, was to allow the blood to flow out before it spoiled. In cases of depressed fractures, there was no need to trephine since there were already passages in the fractured skull for the blood to escape.[17]

By Galen's time (129–199) trephining was in standard use in treating skull fracture for relieving pressure, for gaining access to remove skull fragments that threatened the dura, and, as in Hippocratic medicine, for drainage. Galen discussed the techniques and instruments in detail and advocated practicing on animals, especially the Barbary "ape" (*Macaca sylvana*). He was well aware of avoiding damage or pressure on the dura and indeed carried out experiments on the effect of pressing on the dura in animals.[18]

———

TREPANATION IN ANCIENT CHINA

The possibility that trepanation was practiced in ancient China is suggested by the following story about Cao Cao and Hua Tua, from a historical novel attributed to Luo Guanzhong, written in the Ming dynasty (1368–1644) and set in 168–280 at the end of the Later Han dynasty. Cao Cao was commander of the Han forces and posthumously Emperor of the Wei dynasty, and Hua Tuo was (and still is) a famous physician of the time.

> Cao Cao screamed and awoke, his head throbbing unbearably. Physicians were sought, but none could bring relief. The court officials were depressed. Hua Xin submitted a proposal: "Your highness knows of the marvelous physician Hua Tuo? . . . Your highness should call for him." . . .
>
> Hua Tuo was speedily summoned and ordered to examine the ailing king. "Your Highness's severe headaches are due to a humor that is active. The root cause is in the skull, where trapped air and fluids are building up. Medicine won't do any good. The method I would advise is this: after general anesthesia I will open your skull with a cleaver and remove the excess matter, only then can the root cause be removed." "Are you trying to kill me?" Cao Cao protested angrily . . . [and] . . . ordered Hua Tuo imprisoned and interrogated.
>
> Ten days later Hua Tuo died . . . his medical text was lost upon his death.[19]

WESTERN MEDICINE

From the Renaissance until the beginning of the nineteenth century trephining was widely advocated and practiced for the treatment of head wounds.[20] The most common use was in the treatment of depressed fractures and penetrating head wounds. However, because of the high incidence of mortality particularly when the dura was penetrated, there was considerable debate in the medical literature throughout this long span about if and when to trephine.[21] Besides trephining in cases of skull fracture, the Hippocratic practice of "prophylactic trephination" in the absence of fracture after head injury continued to persist. For example, in the 1800's Cornish miners "insisted on having their skulls bored" after head injuries, even when there was no sign of fracture.[22]

Until the early nineteenth century trephination was done in the home (figures 1.4 and 1.5). However, when the operation was moved to hospitals, the mortality was so high that trephination for any reason including treatment of fractures and other head injury declined precipitously.[23] The practice was so dangerous the first requirement for the operation was said to be "that the wound surgeon himself must have fallen on his head."[24] Or as Sir Astley Cooper put it in 1839, "If you were to trephine you ought to be trephined in turn."[25] It was against this background that the discovery of Neolithic trephining was so unbelievable to the American and French medical communities in the middle of the nineteenth century. Eventually, the introduction of modern antisepsis and prophylaxis of infection at the end of the nineteenth century, as well as an increased understanding of the importance of intracerebral pressure in head injury, allowed trephination to return as a common procedure in the management of head trauma.[26]

In modern neurosurgical practice, trephining is still an important procedure but it is no longer viewed as therapeutic in itself. It may be used for exploratory diagnosis, for relieving intracerebral pressure (as from an epidural or subdural hematoma), for debridement of a penetrating wound, and to gain access to the dura and thence the brain itself (for example, to provide a port through which a stereotactic probe can be introduced into the brain.)

Epilepsy and Mental Disease

In the European medical tradition, in addition to its use in treating head injury, trephining has been an important therapy for two other conditions, epilepsy and mental illness.

———

Figure 1.4
A sixteenth-century woodcut of a trephination in the home. Note the man warming a
cloth dressing, the woman praying, and the cat catching a rat (Dagi, 1997).

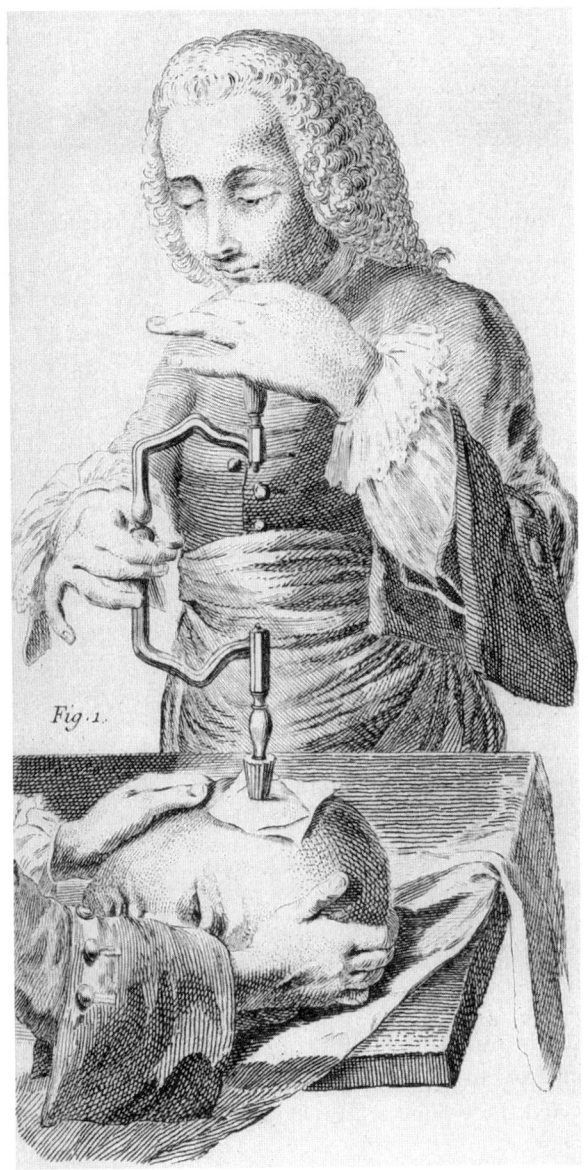

Figure 1.5
Trepanation scene from Diderot's encyclopedia (Diderot and D'Alembert, 1761) Used with kind permission of the Princeton University Rare Books Collection, Princeton, New Jersey.

The tradition of trephining as a treatment for epilepsy begins as early as Aretaeus the Cappadocian (ca. 150), one of the most famous Greek clinicians, and lasted into the eighteenth century. The thirteenth-century surgical text *Quattuor magistri* recommended opening the skulls of epileptics "that the humors and air may go out and evaporate."[27] However, by the seventeenth century trephination for epilepsy was beginning to be viewed as an extreme measure, as in Riverius, *The Practice of Physick* (1655):

> If all means fail the last remedy is to open the fore part of the Skul with a Trepan, at distance from the sutures, that the evil air may breath out. By this means many desperate Epilepsies have been cured, and it may be safely done if the Chyrurgeon be skilful.[28]

By the eighteenth century the incidence of trephining for epilepsy had declined and its rationale changed. Now rather than the idea of allowing an exit for evil vapors and humors, the purpose was to remove some localized pathology. By the nineteenth century trephining for epilepsy was confined to the treatment of traumatic epilepsy, that is, cases associated with known head injury.[29]

Another use of trephining was as a treatment for mental disease. In his *Practica Chirurgiae*, Roger of Parma (ca. 1170) wrote:

> For mania or melancholy a cruciate incision is made in the top of the head and the cranium is penetrated, to permit the noxious material to exhale to the outside. The patient is held in chains and the wound is treated, as above, under treatment of wounds.[30]

Robert Burton, in his *Anatomy of Melancholy* (1652), also advocated boring a cranial hole for madness, as did the great Oxford neuroanatomist and physician Thomas Willis (1621–1675). (See chapter 5.)

Probably the most famous depictions of apparent trephining for mental disease are in early Flemish Renaissance painting. Thus, Hieronymus Bosch's (1450–1515) *The Cure for Madness (or Folly)*, also known as *The Stone Operation* (figure 5.1) shows a surgical incision being made in the scalp. The inscription has been translated in part "Master, dig out the stones of folly."[31] There are similar depictions of the removal of stones from the head by Peter Bruegel, Jan Steen, Pieter Huys, and other artists of the time (figures 5.2, 5.3, and 5.4). The interpretations of these paintings by art historians and their relation to the medical practice of the times are discussed in chapter 5.

By the eighteenth century, "most reputable and enlightened surgeons gave up the practice of . . . [trephination] . . . for psychiatric aberrations or headache without evidence of trauma. Thus, . . . the skull was never to be trephined for 'internal disorders of the head.'"[32]

Trephining in Africa

Herodotus describes the Libyans as cauterizing the heads of their children to "prevent them being plagued in their afterlives by a flow of rheum from the head." And indeed, trephined skulls have been found among the people he was probably writing about, the Tuareg nomads.[33]

A important source of information on the motivations for trephination is contemporary traditional practitioners and their patients. There are literally hundreds of twentieth-century accounts of trephination, particularly in Oceanic and African cultures. Especially detailed and recent ones concern the Kisii of South Nyanza in Kenya and include photographs of the surgical instruments, practitioners, and patients; X-rays of the skulls of surviving patients; detailed interviews; and even a documentary film.[34]

Trephining among the Kisii is carried out primarily for the relief of headache after some kind of head injury. According to Margetts, it is not done for "psychosis, epilepsy, dizziness or spirit possession."[35] The operation is carried out by general practitioners of medicine and takes a few hours. Restraint rather than anesthesia is used. The hole in the skull is usually made by scraping with a sharp knife with a curved tip to avoid injuring the dura. Various medicines are administered before, during, and after surgery but their nature does not seem to have been studied. Mortality, by one authority, is described as "low, perhaps 5 per cent." The practitioners and patients seem to be quite satisfied with the results of the operation.[36]

Although headache after head injury is the most prevalent reason given for trephining by contemporary practitioners of traditional medicine in Africa and elsewhere, other reasons are cited in the literature such as "to let out the evil spirits which were causing an intractable headache."[37]

TREPHINING ON THE INTERNET

Today, the practice of trephining is not confined to surgical suites or traditional medicine men. It is advocated by the International Trepanation Advocacy Group as a means of enlightenment and enhanced consciousness. Their general idea is that when the skull sutures close in childhood it "inhibits brain pulsations causing a loss of dreams, imagination and intense perceptions." Trephining a small hole, they say, "restores the intracranial pulse pressure which leads to a permanent increase of the brain-blood volume which leads to accelerated brain metabolism and more areas of the brain functioning simultaneously" and "increased originality, creativity and...testosterone level."[38] Beyond such "physiological" arguments, the group supports the practice by pointing out its ancient, widespread, and continuing presence in other cultures. This particular form of alternative medicine recently gained considerable if not entirely positive publicity: in November 1998 it was featured on *ER*, the television soap opera set in an emergency ward.

Much of the defense for alternative medicine treatments is that they must work because they have been around for such a long time, an apparently attractive argument for the increasing popularity of five-thousand-plus-year-old Chinese traditional medical practices. However, the case of trephining suggests that just because a procedure is very old does not mean it is necessarily an effective one, at least for enhanced enlightenment and creativity.

Trepanation as an Empirical if Not a Rational Procedure

The most common view of the prehistoric and the non-Western practice of trephining, especially in the absence of a depressed fracture, was that it represented some kind of "superstition," "primitive thinking," "magic," or "exorcism." Yet an examination of the reasons for the practice among the Hippocratic and early European doctors as well as among contemporary Kenyan practitioners suggests a different view. Trephining may have appeared, in these contexts and cultures, to have been an effective empirical approach to head injury and the headaches that often accompany them. Headaches after head injury often do feel like "a pounding" and "pressure" inside the head and thus the idea that a hole in the skull would relieve them is not necessarily magical or bizarre. Furthermore, epidural bleeding does sometimes accompany head injury, and in these cases trephining might have actually reduced intracranial pressure. Finally, the apparently excellent survival rate meant that the procedure, at least until it moved into a hospital setting, may have met the prime requirement of medicine, "do no harm."

––––––

Postscript

The first International Colloquium on Cranial Trepanation in Human History was held at the University of Birmingham in April 2000. Papers from this unique three-day meeting were published as *Trepanation: History, Discovery, Theory*,[39] which provides the most complete review of the subject to date. A major achievement of the meeting was the demonstration that

––––––

trepanation was widespread in many regions of Europe, Asia, Africa, Oceania, and the Americas in both preliterate and literate periods. The volume also contains illustrations of trephined skulls from many cultures and of the great variety of instruments used.

Another interesting development was the return of E. L. Margetts to the Kisii of Kenya, whose trepanning practices he had studied 25 years earlier.[40] He estimates that there may now be more than 100 surgeons carrying out the operation. Unlike in the past, they now use modern Western local anesthetics injected into the scalp prior to surgery. However, the reasons for the very low rate of infections still have not been studied systematically.

Since my original article, there seems to have been an increase in Internet sites advocating trepanning and often self-trepanning for the treatment of, among other disorders, depression, chronic fatigue syndrome, and stress and to improve mental "energy and vigor."[41]

The *British Medical Journal* took these developments seriously enough to issue a warning of their dangers:

> Doctors have warned about the dangers of trepanning after the launch of several websites promoting the "do it yourself" surgery and the case of a Gloucestershire woman who drilled a 2 cm diameter hole in her skull. Concern has been expressed about the growing interest in trepanning for several conditions, including depression and chronic fatigue syndrome. Concern is also growing about the increasing promotion of trepanning, including videos, T-shirts, and a virtual trepanning shopping mall on the internet.[42]

Trepanning received widespread publicity when the surgeon Stephen Maturin carried out the procedure on a sailor in view of the assembled crew in the film *Master and Commander: The Far Side of the World*, based on the Patrick O'Brian naval novels about the Napoleonic Wars.[43]

Notes

Earlier versions of this article were published in *The Neuroscientist* (5: 263–269 [1999], "A hole in the head") and in *Trepanation: History, Discovery, Theory* (ed. R. Arnott, S. Finger, and C. U. M. Smith [Swets and Zeitlinger, 2003], "Trepanation from the Paleolithic to the Internet"). My motivation in choosing this subject was to start with the earliest evidence for knowledge of the brain among people other than the hunter and the cook.

1. Squier, 1877. Squier was an engineer turned journalist with an interest in aboriginal culture. As a reward for helping the presidential campaign of Zachary Taylor he was appointed Charge d'Affaires to the "Republic of Central America," which enabled him to travel extensively and write abut the region (Fernando and Finger, 2003).

2. New York Academy of Medicine, 1865.

3. Lisowski, 1967; Majno, 1975; Bakay, 1985.

4. Schiller, 1992. Schiller claims that Broca was less racist than most of his contemporaries. This may help explain Broca's readiness to accept the trephined skull as a result of skilled surgery by an Inca Indian.

5. Schiller, 1992.

6. Schiller, 1992; Sigerist, 1987.

7. Schiller, 1992; Sigerist, 1987; Finger and Clower, 2003.

8. Lisowski, 1967; Margetts, 1967; Saul and Saul, 1997.

9. Lisowski, 1967; Saul and Saul, 1997.

10. Schiller, 1992.

11. Hippocrates, 1927.

12. Wilkins, 1997; Thompson, 1938.

13. Lisowski, 1967.

14. Bakay, 1985; Wilkins, 1997; Woodall, 1639.

15. Hippocrates, 1927; Rocca, 2003a. A more recent discussion of Hippocratic trephination is Dimopoulous et al., 2008.

16. Hippocrates, 1927.

17. Majno, 1975.

18. Rocca, 2003a; Dimopoulous et al., 2008.

19. Guanzhong, 1991; the real Hua Tuo (ca. 200) had "enduring fame" as a surgeon for using some type of general anesthesia in surgery, for medical gymnastics (*Frolics of the Five Animals*), and for his skill as an acupuncturist (Lu and Needham, 1980). From the Wei dynastic history, Lu and Needham (1980) give a different version of Hua Tuo's interaction with the emperor than the one in Guanzhong's novel, viz., the emperor suffered from migraine, mental disturbances, and dizziness, and Hua Tuo gave acupuncture at a point in the sole of the foot and he was immediately cured. Since Lu and Needham's classic, Hua Tuo has become popular on Chinese medicine Internet sites, there are a number of remedies named after him, and there is even a translation of his supposed work *Classic of the Central Viscera* (Hua Tuo, 1993).

20. Lisowski, 1967; Bakay, 1985; Woodall, 1639; Goodrich, 1997; Dagi, 1997; Wehrli, 1939.

21. Dagi, 1997.

22. Rosen, 1939.

23. Bakay, 1985.

24. Majno, 1975.

25. Flamm, 1997.

26. Dagi, 1997.

27. Aretaeus, 1856; Temkin, 1971.

28. Temkin, 1971.

29. Temkin, 1971.

30. Valenstein, 1997.

31. Cinotti, 1969.

32. Mettler and Mettler, 1945.

33. Margetts, 1967.

34. Margetts, 1967; Grounds, 1958; Coxton, 1962.

35. Margetts, 1967.

36. Margetts, 1967; Grounds, 1958; Coxton, 1962.

37. Margetts, 1967.

38. This account and the quotes were derived from a Web site, www.Trepan.com, accessed in 1999, which no longer exists in its original form. Trepanation sites on the Internet seem to come and go.

39. Arnott, Finger, and Smith, 2003. For some reason, the presentation at the meeting of a modern-day advocate of trepanning for "raising consciousness" was omitted from this volume of the papers from the meeting.

40. Margetts, 1998.

41. Oft-cited print accounts include that of Mitchell, 1999.

42. Dobson, 2000.

43. This incident is from O'Brian (1984).

Heart versus Brain: Galen and the Squealing Pig

Background

Galen of Pergamon (129– ca. 213), "Prince of Physicians," was the out-standing and most influential physician, anatomist, and experimental physi-ologist of classical antiquity. The following essay, "Galen and the Squealing Pig," describes his most famous public physiological demonstration. Galen used this live demonstration to support the still disputed view that the brain, as opposed to the heart, controls thought and action. This new preface to the paper provides the background of the controversy over the role of the heart and the brain.

Many preliterate cultures and the early Egyptian, Mesopotamian, In-dian, and Chinese civilizations attributed psychological functions to the vis-cera rather than the brain. In some of these cultures, such as the Egyptian and Indian, the heart was the center of all sensory, motor, and mental func-tions. In others, psychological functions were distributed among the viscera. For example, in Mesopotamian thought, the heart was the center of the in-tellect, the liver the center of emotions, the stomach of cunning, and the uterus of compassion. In the Chinese *Yellow Emperor's Classic* mental func-tions are also distributed among the internal organs.[1]

The Alcmaeon-Hippocratic-Alexandrian Encephalocentric View

The explicit belief that the brain controlled sensation, cognition, and movement arose among the pre-Socratic philosopher-physicians of the fifth century BCE. The first of these was Alcmaeon of Croton (ca. 450 BCE) who is
said to have been the first to dissect as an intellectual inquiry, to have
described the optic nerves, and to have written:

> The seat of sensations is in the brain. This contains the govern-
> ing faculty. All the senses are connected in some way with the
> brain; consequently they are incapable of action if the brain is
> disturbed or shifts its position, for this stops up the passages
> through which senses act. This power of the brain to synthesize
> sensations makes it also the seat of thought: the storing up of
> perceptions gives memory and belief, and when these are stabi-
> lized you get knowledge.[2]

At about the same time we find the following famous paean to the importance of the brain in the Hippocratic treatise (ca. 425 BCE) *On the Sacred
Disease*,

> It ought to be generally known that the source of our pleasure,
> merriment, laughter, and amusement, as of our grief, pain, anx-
> iety, and tears, is none other than the brain. It is specially the
> organ which enables us to think, see, and hear, and to distin-
> guish the ugly and the beautiful, the bad and the good, pleasant
> and unpleasant.... It is the brain too which is the seat of mad-
> ness and delirium, of the fears and frights which assail us, often
> by night, but sometimes even by day; it is there where lies the
> cause of insomnia and sleep-walking, of thoughts that will not
> come, forgotten duties, and eccentricities.[3]

The emphasis on the brain in sensation and thought was further developed by the Alexandrian anatomists Herophilus and Erasistratus (3rd C. BCE) who carried out the first systematic and detailed studies on the anatomy of the brain including of humans, probably some of them still alive.[4] Herophilus and Erasistratus worked at The Museum in Alexandria founded by Ptolemy I (367–283 BCE), Alexander's friend and general and the first Greek ruler of Egypt, who, as a young man, had been tutored, along with Alexander, by Aristotle.

Aristotle's Cardiocentric View

The idea that the brain is central for sensation, movement, and mentation was a dominant tradition in Greek medicine from Alcmaeon through the Hippocratics and Alexandrians to Galen. However there was an opposing tradition in Greek philosophy, beginning with Aristotle, that held that the heart—not the brain—was the "command center" (*hegemonikon*) of the soul, the center of sensation, movement, and cognition.

Aristotle (384–323) knew the arguments for the hegemony of the brain of Alcmaeon, the Hippocratic school, and others and argued against them in detail.[5] In support of his cardiocentric view, Aristotle adduced several lines of evidence including (a) *anatomical*—the heart connects with all the sense organs but the brain does not (on dissection, blood vessels are indeed more prominent than nerves); the heart is centrally placed whereas the brain is peripherally located; (b) *embryological*—the heart develops before the brain; (c) *comparative*—all animals have a heart, but invertebrates, which do have sensation, have no brain; (d) *observational*—the heart, unlike the brain, is sensitive to touch; the heart but not the brain is affected by emotions; and (e) *physiological*—the heart provides blood needed for sensation but the brain is bloodless, without sensation; the heart is warm like higher life but the brain is cold; the heart but not the brain is essential for life.

But there was an essential approach missing, namely, the clinical approach, the study of the brain-injured human. The champions of the hegemony of the brain, Alcmaeon, Hippocrates, Herophilus, and Erasistratus, were all practicing physicians. The evidence they had given in support of their opinions was strictly clinical. Since there is no evidence of systematic experiments on the brain and nervous system until Galen in the second century, the accidents of nature were the only sources of information about the functions of the brain. It is hard to conceive of Aristotle, in the course of his strictly zoological observations and dissections, coming across evidence strongly contradicting his view of the brain and heart.

Although he came from a family of physicians and had been slated for a medical career, Aristotle at no time seemed interested in medicine or medical writing. Indeed, medicine appears to be one of the few things that this universal genius was not interested in. However, in the fourth century BCE, the study of the effects of damage to the human brain was the most likely way of getting a "more correct" view of the brain than Aristotle had.[6]

The Tripartite Soul: Democritus and Plato

Democritus (460–371 BCE) the pre-Socratic famous for his atomic theory and a friend of Hippocrates, took an intermediate position on the role of the brain and heart: he distributed the soul and mental function over three regions.[7] The brain was the central organ of consciousness and thought, the heart the center of emotion, and the liver the site of lusts and appetites. Plato (428–347 BCE), following Democritus, put reason or intellect, the highest and immortal part of the soul, in the brain, which controls the rest of the body. In the *Timaeus* he wrote, "It is the divinest part of us and lords over all the rest."[8] Sensations were integrated in the heart, and the liver was responsible for lusts.

The Stoics

The Stoic philosophers adopted Aristotle's cardiocentric view of all mental function. This school, founded by Zeno (344–262 BCE), became widely influential through the teachings of Chyrsippus (280–207 BCE). The Stoic school strongly objected to Plato's separation of reason, emotion, and desire and argued for the unification of all aspects of the human mind and soul in one location.[9] The location they chose was the heart. In addition to Aristotle's arguments for the hegemony of the heart over the brain, the Stoic philosophers stressed the association of thought with speech and in turn with breath, which was considered a cardiac function. Here is their argument in a nutshell:

> Speech passes through the windpipe. If it were passing from the brain it would not pass through the windpipe. Speech passes from the same region as discourse. Discourse passes from the mind. Therefore the mind is not in the brain.[10]

The Stoics developed an elaborate philosophy unifying ethics and politics with a view of the laws of the universe that provided a guide to human happiness. This system evolved over several hundred years, but only two aspects concern us here. First, it was by far the dominant philosophical school in Rome in Galen's time. Indeed, Galen's most famous patient was the Emperor Marcus Aurelius (121–180), a distinguished Stoic author in his own right. Second, it rejected the medical tradition from Alcmaeon and Hippocrates through the Alexandrians that sensation and thought were functions of the brain. Rather, it unified all mental functions in the heart.[11]

Galen was a contentious and combative writer on many medical and philosophical subjects, but perhaps his major battle was with the Stoic cardiocentric view of mind. He not only wrote a long and detailed attack on it but also carried out what he considered a public experimental refutation of it, the subject of the following essay.[12] This demonstration on a large pig became one of the most famous public physiology demonstrations of all time.

GALEN AND THE SQUEALING PIG

Galen, who lived in the Roman Empire in the second century, was the greatest experimental physiologist and anatomist of classical antiquity (figure 2.1). He was the most important figure in classical medical science and represents the peak of ancient Western anatomy, physiology, and medicine. His ideas were so pervasive that the medieval world saw the structure and function of the human body largely through his eyes.[13] Today, his extensive writings (see box 2.1) provide a vivid account of the context, controversies, and achievements of the 600 years of classical biology and medicine. After a brief account of Galen's life and his contributions to neuroscience, we consider his most famous experiment on the nervous system, namely, the effect of cutting the recurrent laryngeal nerves. This demonstration, carried out with a squealing pig as subject, was famous in its own time and for centuries later. Although there was a long tradition before Galen that the brain mediated sensation, cognition, and movement, the contrary view—that the heart subserved these functions—was dominant in Galen's time. Galen's findings on the recurrent laryngeals were viewed as strong experimental evidence for the primacy of the brain in behavior and thought.

Life

Galen was born of upper-class parents in 129 in Pergamon, a rich and ancient Greek city located on the site of the present-day Turkish city of Bergama, about 15 miles from the Ionian Sea. It was a traditional rival of Alexandria, and the rulers of Egypt had banned the export of papyrus to Pergamon in order to block the development of a rival library there. In response, the Pergamenians developed a new writing material made from animal skins called *charta pergamena*, which gives us the English *parchment*. In Galen's time, both cities were part of the Roman Empire, then at its peak.[14]

   Pergamon was the site of one of the most famous Aesclepieia, named after the Greek god of medicine Aesclepius. It was a combination of a

Figure 2.1
Portrait of Galen. From the Juliana Anicia Manuscript, written in 487 (Singer, 1957).

Box 2.1
Galen's Writings

Galen published voluminously on almost every branch of medical science and
medical practice known in his time as well as on philosophy, rhetoric, and his
own life story (Sarton, 1954). Most of Galen's works are lost; in the 1820s the
surviving Greek works were collected and translated into Latin by K. G. Kuhn.
They make up 22 very large volumes. There are also some other works which survived only in Arabic. Much of Galen's writings remain untranslated into English,
and most of the English translations were published only in the last few decades.

Galen discusses the circumstances surrounding his first public demonstration
of the recurrent laryngeal nerves in *On Prognosis*. He gives an account of the
recurrent laryngeal in *On the Usefulness of Parts of the Body*, but the most detailed
account is in the later books of *On Anatomical Procedures* (or *De Anat. Admin.*;
Galen's works are most commonly denoted by abbreviations of their titles in
Renaissance Latin).

A brief history of *On Anatomical Procedures* gives the flavor of the adven-
tures of the ancient texts that managed to survive. It was originally a two-chapter
work written in 169. Soon after, Galen's copies were lost in a fire and others were
unavailable, so Galen eventually wrote a much expanded version in 177. Of this
text, Books I–XI were published but Books XII–XV were destroyed, along with
many of his other works, in another fire and had to be rewritten. Only one manuscript copy survived in Western Europe and it broke off in the middle of Book IX.
It was first printed in 1525 by Aldus in Venice.

In 1844 the remaining books of *On Anatomical Procedures* were discovered as
an Arabic manuscript in Oxford's Bodleian library. They had been translated from
Greek into Syriac by the great Arab physician Hunain ibn Ishaq (809–873) and
then into Arabic by him and his nephew. The Bodleian Arabic manuscript was
translated into German in 1906 and then into English in 1962 by W. L. H. Duckworth with the help of an anatomist and an Arabist.

---

medical treatment center, a healing temple, a pilgrimage site, and a medical school. A common therapeutic technique was to induce sleep with drugs ("incubation") and then to whisper in the sleeping patient's ear that a particular treatment would be efficacious for his ills. The next day, when the patient was told the treatment plan by his doctor, he interpreted his dream as having foretold his doctor's plans. The collection of buildings that made up the Pergamene Aesclepium, along with monuments left by rich patients in recognition of their cures, can still be visited today.[15] Pergamon was also the site of a major gladiatorial school.

Galen's father, an intellectually inclined architect, began to tutor his son in philosophy and mathematics at an early age. However, when Galen was 16 his father had a dream, supposedly sent by Aesclepius, that led Galen to begin the study of medicine. During his four years as a medical student in Pergamon, he published three medical texts, one on the uterus, one on the eye (now lost), and one on medical methodology. Over the next eight years he continued his medical studies at three other major medical centers including Alexandria, the leading center of medical research and teaching. In this period, he acted more as a postdoc than a medical student: carrying out research, writing on a variety of medical subjects, and maintaining his interest in philosophy. Finally, at the age of 28 he returned to Pergamon and was appointed physician to the gladiators, a rich source of clinical and anatomical material, as most combats ended in serious or fatal injury.[16]

In 161, a war between Pergamon and its neighbors caused the gladiatorial "games" to be closed, so Galen set off for the Rome of Marcus Aurelius.[17] In Rome at this time, there was an unusually close relationship between, on one hand, the ruling politicians and aristocrats and, on the other, the intelligentsia, particularly the philosophers and rhetoricians known as Sophists, but including other philosophers, physicians, and scientists. In this period, known as the "Second Sophistic," the Empress Julia hosted a salon of philosophers and writers; Greek intellectuals married Roman aristocrats; and the rulers of the empire sponsored and attended scientific demonstrations and lectures.[18]

In this unusual environment, Galen rapidly rose to the highest social and professional level of Roman society. Helped by curing several influential political figures, he built up a large and successful practice. One of his powerful patrons was Flavius Boethus, a former consul and later governor of Syria. Boethus encouraged him to compose his first major anatomical and physiological works. He also arranged for Galen to present a series of public anatomical demonstrations and lectures which were well attended by the intellectual and political elite. The most famous of these was Galen's demonstration of the functions of the recurrent laryngeal nerves, discussed in detail below. During this period Galen also became involved in several acrimonious disputes with other leading physicians.[19]

Perhaps because of these disputes, or because of an epidemic in Rome, Galen returned to Pergamon. Soon after, the co-emperors Marcus Aurelius and Lucius Verus recalled him to accompany their troops in the field. He talked his way out of this assignment by becoming personal physician to Marcus's son Commodus. He continued to serve Commodus when he became emperor and treated the subsequent emperor Septimus Severus as well. Galen repeatedly boasted that as the result of his professional accomplishments he became well known to all the leading philosophers and writers of his time as well as all the emperors. He continued to treat patients, research, write, quarrel with other physicians, and be lionized in high society until his death in about 213.[20]

Physiological System

Galen believed that physiology and anatomy formed the critical bases of medical practice and he wrote extensively on both subjects. His physiological system totally dominated physiology and medicine until William Harvey in the sixteenth century and continued to be very influential until the nineteenth century.[21]

In his system the fundamental principle of life was pneuma (akin to the *chi* of Chinese medicine and the *vayu* of Indian medicine.) It entered the body from the all-pervading world spirit during breathing and passed to the lungs and then via the pulmonary vein to the left ventricle where it mixed with the blood. The blood had been made in the liver from chyle brought from the intestines by the portal vein. The liver had also given the blood the lowest type of pneuma, natural spirits, which was believed to be innate in all living tissue. The blood with its natural spirits and nutritive material was now distributed throughout the body by the veins in a tidal, or ebbing and flowing, motion. Some of the blood entered the right side of the heart, from which it had two possible routes. Most of it stayed in the right ventricle, from which its impurities were carried off by the pulmonary artery to the lungs and exhaled. A smaller portion of it trickled into the left ventricle through the holes that Galen thought existed in the interventricular septum. There it mixed with air, which had come in from the lungs via the pulmonary vein, and thereby became transformed into a higher type of pneuma, vital spirits. The vital spirits were distributed to the body and head via the blood in the arteries. Some blood carrying the vital spirits went to the base of the brain to the "rete mirabile" (a network of blood vessels at the base of the brain found only in the ox and some other animals, but not in humans—although it was described and drawn as a very prominent feature of the human brain until Vesalius). Both here and in the choroid plexi inside of the ventricles, vital spirits became transformed into the highest pneuma, animal or psychic spirits. The brain ventricles were an important storage site for this psychic pneuma; from there it was distributed throughout the brain and via the (hollow) nerves to the rest of the body. All this remained dogma for over 1,500 years.[22]

Neuroscience Achievements

Whatever the weaknesses, from our point of view, in some of Galen's theoretical views, he made a number of major discoveries, particularly on the anatomy and physiology of the nervous system. He described in detail the course of nine, if not ten, of the cranial nerves (although he grouped them as seven pairs), as well as the sympathetic nerve trunks.[23] He distinguished sensory and motor nerves for the first time and thought that this distinction derived from their source in the brain, a clear statement of Müller's doctrine of specific nerve energies. Galen's descriptions of the gross anatomy of the brain were very accurate, particularly with respect to the ventricles and the cerebral circulation, both important in his physiological system. Galen usually presented his dissections as if they were of the human, but, in fact, they were invariably of animals, usually the ox in the case of brain anatomy, the Barbary "ape" (the macaque *M. sylvana*) for cranial nerve anatomy, and pigs for vivisection. It was only very recently, when Galen's descriptions were evaluated in terms of the actual species dissected, that their great accuracy was recognized.[24]

Galen was the first to carry out systematic experiments on the effects of experimental lesions of the nervous system. Before him, no one had, as he put it, "ever taken the trouble to make a section themselves or put a ligature around parts in the living animal in order to learn which function is injured." He usually used a pig in these experiments "to avoid seeing the unpleasant expression of the ape when it is vivisected."[25]

Galen realized that the spinal cord was an extension of the brain. In his brilliant and systematic experiments on sectioning the cord he compared the effects of hemi- and total transection at different levels and noted that injuries interfered with sensory and motor function below the level of the section; that hemisection affected only one side; and that sagittal section did not produce paralysis. He accurately described the different roles of the spinal nerves in respiration. He even came very close to the Law of Spinal Roots: "The physicians do not even know that there is a special root at the origin of the nerves which are distributed to the entire hand and from which sensation arises; [nor do they know] that there is another root for the nerves moving the muscles."[26]

Galen used piglets in his experiments on brain lesions. He found that anterior brain damage had less deleterious effects than posterior.[27] He viewed sensation as a central process since he knew from his clinical observations and animal experiments that sensation could be impaired by brain injury even when the sense organs were intact. Since animals could survive lesions that penetrated to the ventricles, Galen thought the soul was not located there but rather in the cerebral substance.[28] He ridiculed the view of the Alexandrian anatomist Erasistratus that intelligence was correlated with the number of cerebral convolutions, noting, "Even donkeys have a complex encephalon, whereas judging by their stupidity ought to be perfectly simple and uncomplicated."[29] This view continued to be cited in denigrating the role of the cerebral cortex well into the eighteenth century.[30]

One way to measure Galen's achievements is to note how long it took for them to be superseded. His neuroanatomical discoveries were not surpassed until Vesalius in the seventeenth century, his studies of spinal transection not until Magendie in the eighteenth century, and his work on hemispheric function not until Gall in the nineteenth century.

Brain and Heart

The Hippocratic writers (4th C. BCE), particularly the author of *On the Sacred Disease*, had vigorously championed the hegemonic role of the brain in sensation, movement, and thought as did Plato (4th C. BCE). Indeed this view had been held earlier by several of the pre-Socratic philosopher physicians starting with Alcmaeon of Croton (6th C. BCE). Furthermore, in the second century BCE, the Alexandrian anatomists Herophilus and Erasistratus had explored the structure of the brain and its role in sensation, movement, and mentation in detail and probably in living humans.[31]

Yet, 400 years later in the Rome of Galen's time, it was the common belief, even among leading physicians, that the heart rather than the brain was the central organ of sensation, movement, and mentation. This had been Aristotle's view much earlier (4th C. BCE) and was a central tenet of the Stoic philosopher Chrysippus (280–207 BCE), whose views continued to be highly influential, if not dominant, in Galen's time.[32] Indeed

Chrysippus's importance is indicated by the large amount of space Galen devotes to refuting his views in *On the Opinions of Hippocrates and Plato* and other works.[33]

The Squealing Pig Demonstration

Before coming to Rome, Galen had conducted extensive experiments on the nerves that control breathing. In the course of one of these experiments, carried out on a strapped down pig as it struggled and squealed, Galen accidentally cut the recurrent laryngeal nerves which innervate the larynx and the pig stopped squealing but not struggling. Subsequently, he traced out the course of these nerves in detail in a variety of animals including long-necked birds. He accurately described how the nerves begin as a branch of the Vagus (his "6th" cranial nerve), extend down far past the larynx and then loop around the aorta on the left and the subclavian artery on the right before traveling upward to the larynx (figure 2.2). He also confirmed that their bilateral section in dogs, goats, bears, lions, cows, monkeys and other animals eliminated vocalization. Furthermore, he mentions two instances of loss of vocalization in human infants following accidental injury to the recurrent laryngeals in the course of surgery to remove goiters.[34]

In Rome, his powerful patron Boethus arranged for him to conduct a public demonstration that section of the recurrent laryngeal nerves would eliminate vocalization in the pig. Boethus hired the hall, obtained the requested pigs and widely advertised the event. Many distinguished politicians and scholars came, including Alexander Damascenus, an Aristotelian philosopher (see figure 2.3). Before beginning the surgery, Galen gave a brief summary of the demonstration and noted that "there is a hairlike pair [of nerves] in the muscles of the larynx on both left and right, which if ligated or cut render the animal speechless without damaging either its life or functional activity." At that point Alexander Damascenus interrupted, objecting, "Even if we are shown that section of these nerves in animals renders them mute, it is not necessary to believe it true of human beings."

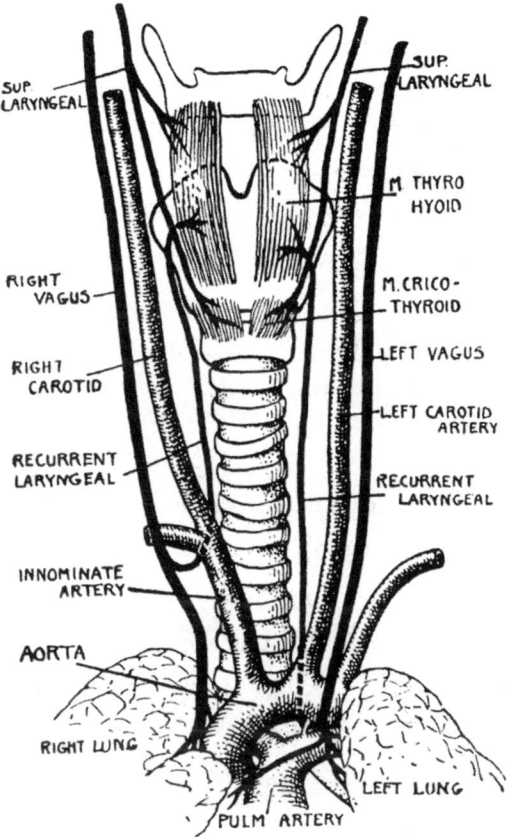

Figure 2.2
The U-shaped course of the recurrent laryngeal nerves (after Singer, 1957). The laryngeal nerves are branches of the vagus nerve that start far below the larynx and then loop under the subclavian artery on the right and under the aorta on the left before moving upward to innervate the larynx.

Figure 2.3

Galen demonstrating the effect of cutting the recurrent laryngeal nerves in a pig. Labeled in the audience are Barbarus, former consul and uncle of the co-emperor Lucius Verus; Severus, another former consul, who introduced Galen to Marcus Aurelius; Boethus, who sponsored the demonstration; Paulus, a leading lawyer; Hadrian, a Sophist who held the imperial chair of rhetoric; Demetrius, another Sophist; and Alexander of Damascus. The latter, an Aristotelian and public professor of philosophy in Athens, heckled Galen, thereby causing a postponement of the demonstration. From the bottom panel of the title page of the 1541 Junta edition of *Galen's Works*, courtesy of Yale University, Harvey Cushing/John Hay Whitney Medical Library.

In any case, he said, he would not believe such a demonstration. His insistent view reflected both a general skepticism of the value of sensory information as opposed to logic and geometry in establishing proof and the Aristotelian (and Stoic) belief that thinking and therefore vocalizing were controlled by the heart and not the brain.

Galen later recorded his response to Alexander's interruption:

> When I heard this, I left them and went off, saying only that I was mistaken in not realizing that I was coming to meet boorish skeptics; otherwise I should not have come. . . . On my departure, the others condemned Alexander. . . . When this was made known to all the intellectuals living in Rome . . . they all roundly reproached him and demanded that the dissections be performed in their presence when they assembled all others distinguished in medicine and philosophy. The meeting lasted several days, in which I showed them [the nerves controlling breathing] and how damage to the nerves activating the muscles of the larynx results in a loss of voice. My detractors were all put to confusion when I showed them this and Boethus begged me to give him my lecture notes on it. He even sent trained shorthand writers to whom I dictated all my demonstrations and arguments.[35]

From this shorthand came the treatise *On the Voice*, now lost but probably the basis of the section in *On the Usefulness of Parts of the Body* on the recurrent laryngeals, since that section is very much in the form of a public lecture.[36]

Reasons for the Recurrent Course

Galen's adoption of Aristotle's biological teleology was consistent and relentless. Like Aristotle, he believed everything in the body was designed in

41

the best possible way for the best possible reason, and that the perfection of this design demonstrated the existence and genius of the Creator. (Although believing in a supreme being, Galen had so little interest in Christianity that he did not even distinguish it from Judaism.[37])

What then was the teleological reason for the tortuous route traveled by the laryngeal nerves down from the brain into the trunk and then around the aorta and subclavian arteries and back up to the laryngeal muscles? Before considering Galen's answer, we must summarize his notions of muscular action. Galen thought that when a motor nerve enters the origin of the muscle it breaks up into small fibers, and that the ligaments do so as well. At the point of insertion these muscle and ligament fibers unite to form the tendon, which is then much stronger than if it had been made by the nerves alone. It is this tendon that contracts, while the body of the muscle is mere "flesh" that protects the fibers.[38] Since in muscle action the pulling is done by the tendon (made up in part by the nerve), the nerve must enter the muscle in the direction of pull. This provided a rationale for the recurrent pathway of the laryngeals; they had to enter the larynx muscles from the bottom. Galen compares the passing of the nerves around the aorta and subclavian to a rope passed through a pulley.[39] Apparently, pulley action was not well known to his audience, so he explains it in reference to the "glossocomion," a gadget for reducing fractures of the femur and tibia, which was "a common device familiar to most physicians" and which involved pulley action (see figure 2.4).[40]

> Now if the heart were the source of the nerves, as some think who know nothing of what is to be seen in dissection, it would readily move the [laryngeal] muscles by sending nerves directly into them.... Actually... every nerve obviously takes origin from the brain or spinal cord... [but the laryngeal] muscles could receive nerves from the brain only if they followed a reversed route.... [He then digresses to attack several rivals who fail to recognize the wisdom of nature]. Hence I find it

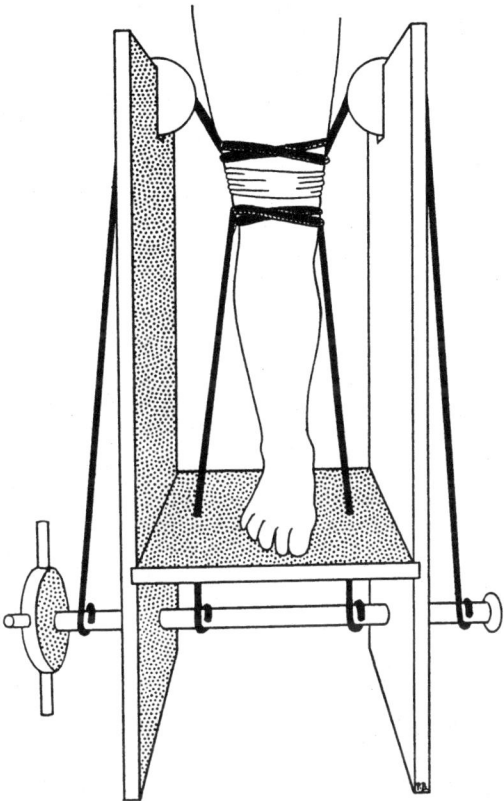

Figure 2.4
Glossocomion for reducing fractures. When the wheel is turned, traction is applied to the leg (Galen, 1968).

necessary to tell now about the devices Nature has employed in order to give the muscles in question their share of nerves and motion. In order that my discourse may be clear you must first understand this reversed motion which is made use of in many devices both by engineers and surgeons.[41]

A long discussion of pulley action in the glossocomion is then followed by a detailed and accurate account of the course of the vagus nerve and its branches, particularly the recurrent laryngeals and how they loop abound the subclavian and aorta. Galen explains that they do not use the clavicle as a pulley, although it is encountered first, because that would make the nerves too "exposed close under the skin and easily harmed by every mischance." Rather the nerves continue down and then use the large horizontal blood vessels as a "pulley or turning post." This account is interrupted several times by exhortations to the audience to pay attention, praise of the Creator, and insults to rival anatomists as in the following passage.

I want you now to pay me closer attention than you would if you were being initiated into the mysteries of Eleusis or Samothrace or some other sacred site . . . You should consider that this mystery is in no way inferior to those and no less able to show forth the wisdom, foresight and power of the Creator of animals, and in particular you should realize that I was the very first to discover this mystery which I now practice. Certainly, no other anatomist has known about any of these nerves or about the things of which I have spoken earlier in the construction of the larynx and this is the reason why they have erred so greatly in determining actions and have not told a tenth of the utilities of the parts. Accordingly, . . . fix your mind now on holier things . . . and follow closely my discourse as it explains the wonderful mysteries of Nature.[42]

Fame of the Squealing Pig Experiment

Galen's experiment on eliminating vocalization by cutting the recurrent laryngeals has been heralded by one twentieth-century student of Galen as "establishing for all time that the brain is the organ of thought and one of the most important additions to anatomy and physiology...as great as the discovery of the circulation of the blood."[43] This and similar claims are somewhat exaggerated and uninformed. As noted above, that the brain was the organ of thought was clearly set out in many pre-Galenic writings and particularly in the Hippocratic treatise *On the Sacred Disease* and in Plato's *Timaeus*. Even Galen, in his lengthy arguments against the Stoic Chrysippus, concentrates on experiments involving direct damage to the brain.[44] It is true, however, that this was the first experimental and publicly repeatable evidence that the brain controlled behavior. Previous evidence for the importance of the brain had been either exclusively clinical or based on indirect inferences from anatomy.

Galen's demonstration on the squealing pig became one of the most famous single physiological demonstrations of all time. It inspired Leonardo da Vinci to produce a beautiful drawing of the recurrent laryngeal nerves.[45] Vesalius gives it a prominent place (without mentioning Galen) in the last chapter of his great *On the Fabric of the Human Body* and included it in his public lectures in Padua.[46] Finally, Renaissance editions of Galen included in their frontispiece an illustration of Galen cutting the recurrent nerve in a huge pig (figure 2.3).

The Persistence of Galenism

At about the time of Galen's death, classical science and medicine die. People prefer to believe rather than to discuss, critical faculty gives way to dogma, interest in this world declines in favor of the world to come, and worldly remedies are replaced by prayer and exorcism.

The worldview of medieval Christendom found Galen's teleology congenial to its own and by a smothering of critical facility froze Galen and all biology into a sterile system for over 1,500 years. Galen was not to blame. Rather than develop his discoveries and methods, the medieval world chose to accept as fixed and unchangeable fact his views in every branch of medicine.

––––––

POSTSCRIPT

Since this article was published in 1998 there have been dozens of new translations, symposia, and scholarly articles on Galen. Indeed, Nutton estimates that "since 1945 a new fragment of Galen and often a whole work has been announced or published every two years."[47] Even his life span has been increased, his previous age apparently based on mistranslations from Arabic.[48] Perhaps the most interesting new work on Galen from a neuroscience perspective is Rocca's *Galen on the Brain*, published in 2003.[49] Although Galen dissected the brains of a great variety of animals including monkeys, his detailed brain anatomy and his model for the human brain was based primarily on the ox brain. Rocca himself followed Galen's descriptions of his brain dissections with actual ox brains and shows how amazingly accurate Galen was, particularly given that Galen's brains were not fixed. He shows how Galen's detailed account of the complex ventricular system was important because of the crucial role it played in his physiological theory. Although Galen placed thought, memory, and the control of movement in the solid portions of the brain and not in the ventricles, his emphasis on the importance of the ventricles contributed to the ventricular (or "cellular") localization of psychological functions by Nemius of Emesa (ca. 400) that dominated medieval thought and was influential into the nineteenth century.[50]

The debate between brain and heart as the seat of intellect continued into the Renaissance. One resolution was to combine the two views. For example, the Arab Aristotelian and physician Ibn Sina (980–1037, known in the west as Avicenna) located sensation, cognition, and movement in the

––––––

brain, which in turn he believed was controlled by the heart.[51] Similarly, according to the thirteenth-century Hebrew encyclopedist Rabbi Gershon ben Schlomoh d'Arles, the brain and heart share functions so "when one...is missing, the other alone continues its activities...by virtue of their partnership."[52] As Scheherazade tells it on the 439th night, when the Caliph's savant asks the brilliant slave girl Tawaddud, "where is the seat of understanding," she answers, "Allah casteth it in the heart whence its illustrious beams ascend to the brain and there become fixed."[53] Eventually, the "debate" became more of a metaphorical one, as Portia's song in *The Merchant of Venice* asks,

> Tell me where is fancies bred,
> Or in the heart or in the head.

And, metaphorically the heart usually wins, as in Longfellow's poem "The Building of the Ship":

> It is the heart and not the brain
> That to the highest doth attain,
> And he who followeth Love's behest
> Far excelleth all the rest.

NOTES

This chapter is an article that appeared in *The Neuroscientist* (4: 216–221 [1998], "Galen and the Squealing Pig"), with a new introduction and postscript. The very first version of this paper was in an early draft of my 1961 doctoral dissertation on the frontal cortex in monkeys.

1. Sigerist, 1951, 1961; Sarton, 1959; Keele, 1957. *The Yellow Emperor's Classic* is now believed to be a compendium of different early Chinese medical writings. See the new introduction by Lo in Lu and Needham (1980).

2. Lloyd, 1975; Longrigg, 1993; Gross, 1998a. All of the writings of the pre-Socratic philosophers are lost. All we know about them comes from later classical writers. These

"fragments" or putative quotes were collected by the ancient so-called doxographers and were assembled by H. Diels in the nineteenth century and translated into English by Freeman (1954).

3. Hippocrates, 1950. Although Hippocrates was a historical figure, it is unclear which of the Hippocratic corpus he wrote, which were written by his followers, which were just found in the library of his medical school on the Aegean island of Cos, or even which were assembled later (Smith, 1979; Lloyd, 1978). *On the Sacred Disease* is usually judged as by, or representative of, Hippocrates and his close followers.

4. Von Staden, 1989; Dobson, 1926–1927, Gross 1998a.

5. Gross, 1995.

6. Gross, 1995.

7. Lloyd, 1975; Longrigg, 1993; Gross, 1998a; Freeman, 1954.

8. Plato, 1920.

9. Tielman, 1996, 2002.

10. Tielman, 1996, 2002.

11. Tielman, 1996, 2002; Bowersock, 1969.

12. Nutton, 1984; Wilson, 1972; Sarton, 1959; Galen, 1978–1984.

13. Sarton, 1954.

14. Sarton, 1954; Wilson, 1972.

15. Sigerist, 1961; Guthrie, 1945; Walsh, 1934, 1935, 1936, 1937, 1939.

16. Sarton, 1954; Wilson, 1972; Walsh, 1934, 1935, 1936, 1937, 1939. Galen (1988), in a work relatively recently translated from Arabic, *On the Examinations by which the Best Physicians Are Recognized*, described the exam he took to get the position at the gladiator school in Pergamon: "A high priest followed this method (of choosing physicians) when I returned to our city from places which I had set out to visit. Although I had not yet completed thirty years of my age he entrusted me with the treatment of all the wounded (men) among those who had fought duels in combat. . . . I performed many anatomical

demonstrations before the spectators: I made an incision in the abdomen of an ape and exposed its intestines: then called upon the physicians who were present to replace them back and to make the necessary abdominal sutures—but none of them dared to do this. We ourselves then treated the ape, displaying our skill, manual training, and dexterity. Furthermore, we deliberately severed many large veins, thus allowing the blood to run freely, and called upon the Elders of the physicians to provide treatment, but they had nothing to offer. We then provided treatment, making it clear to the intellectuals who were present that those physicians who possessed skills like mine should be in charge of the wounded. That man was delighted when he put me in charge of the wounded.... With the exception of two, none of the wounded in my charge died, whereas sixteen individuals had died under my predecessor. Later, another high priest put me in charge of the wounded and in doing so he was even more fortunate. None of the patients under my care died even though each suffered grave and multiple wounds." This was actually written much later in Galen's life than when he got the position in Pergamon and is typical of his grandiose and inflated self-advertisement.

17. Sarton, 1954.

18. Bowersock, 1969.

19. Bowersock, 1969; Sarton, 1954; Wilson, 1972; Nutton, 1984.

20. Sarton, 1954; Wilson, 1972; Walsh, 1934, 1935, 1936, 1937, 1939; Nutton, 1984.

21. Singer, 1957; Temkin, 1973.

22. Singer, 1957; Temkin, 1973; Siegel, 1968.

23. Galen, 1962; Goss, 1966; Smith, 1971.

24. Singer, 1957; Smith, 1971; Woolam, 1958; Rocca, 2003b.

25. Galen, 1968; Temkin, 1973.

26. Galen, 1962; Goss, 1966; Spillane, 1981; Siegel, 1973.

27. Galen, 1962.

28. Galen, 1978–1984.

29. Galen, 1968.

30. Gross, 1998a.

31. Lloyd, 1975; Longrigg, 1993; Gross, 1998a; Freeman, 1954; Hippocrates, 1950; von Staden, 1989; Dobson, 1926–1927; Gross 1995, 1998a.

32. Plato, 1920; Tielman, 1996, 2002.

33. Galen, 1978–1984.

34. Galen, 1979, 1968, 1962; Farrington, 1932.

35. Galen, 1979.

36. Galen, 1968.

37. Walzer, 1929.

38. Bastholm, 1950.

39. Galen, 1968.

40. Galen, 1968.

41. Galen, 1968.

42. Galen, 1968.

43. Walsh, 1926.

44. Galen, 1978–1984.

45. Gross, 1997b, figure 7.

46. Farrington, 1932.

47. Nutton, 2002.

48. Nutton, 1995.

49. Rocca, 2003.

50. Gross, 1995.

51. Avicenna, 1930.

52. Schlomoh, 1953.

53. Burton, 1885. The colorful Captain Sir Richard Burton himself. Given its racy nature this translation, *The Book of the Thousand Nights and a Night*, was a privately printed edition available only by subscription. Some of his other translations of Eastern erotica, such as *The Scented Garden*, which Burton called "a manual of erotology," were burned by his widow (Lovell, 1998). The later work should be distinguished from his translation of another erotic classic, *The Perfumed Garden* (Burton, 1886), that was again published privately.

The Fire That Comes from the Eye

One of the earliest ideas about vision is that it depends on light that streams out of the eye and detects surrounding objects. This view was attacked in its own time and finally disproved over two thousand years later. Yet the idea of a beam leaving the eye persisted in beliefs both about the evil eye and the power of a lover's gaze. It is still widely held among U.S. schoolchildren and adults. We consider the history and ramifications of ideas about the sources of vision.

Vision: Out of the Eye—Extramission Theories

One of the first neuroscientists we know of was the pre-Socratic Alcmaeon of Croton (ca. 450 BCE). He was the first to advocate the brain as the seat of sensation and cognition and the first to dissect parts of the visual system.[1] Presumably after observing phosphenes resulting from a blow to the head, he noted, "The eye obviously has fire within it, for when one is struck this fire flashes out. Vision is due to the gleaming."[2]

This idea of vision depending on the "fire in the eye" was elaborated by Plato (427–347 BCE) in his cosmological (and rather anti-science) dialogue the *Timaeus*, which was enormously influential in the Middle Ages and beyond.[3] Plato argued that visual fire streams out of the eye and combines with daylight to form a "single homogeneous body" which serves as an instrument for detecting and reporting visual objects:

> Such fire as has the property, not of burning, but of yielding a gentle light, they [the Gods] contrived should become the proper body of each day. For the pure fire within us is akin to this, and they caused it to flow through the eyes . . . Accordingly, whenever there is daylight round about, the visual current issues forth, like to like, and coalesces with the daylight and is formed into a single homogeneous body in a direct line with the eyes, in whatever quarter the stream issuing from within strikes upon any object it encounters outside. So the whole . . . is similarly affected and passes on the motions of anything it comes in contact with . . . throughout the whole body, to the soul, and thus causes the sensation we call seeing.[4]

Theories of vision such as this one, which depend on something streaming out of the eye, are known as extramission theories. Later, the great mathematician Euclid (ca. 300 BCE), in his *Optika*, developed a rigorously and narrowly geometric extramission theory. In this theory, "Rectilinear rays proceeding from the eye diverge infinitely [and] those things are seen upon which the visual rays fall and those things are not seen upon which the visual rays do not fall."[5]

The astronomer and mathematician Ptolemy (127–148) carried Euclid's extramission ideas further and combined them with Galen's (129–ca. 213) work on the anatomy of the eye. Whereas Euclid had postulated discrete rays leaving the eye that became separated with increasing distance, Ptolemy argued that the visual rays formed a continuous bundle or cone.[6]

VISION: INTO THE EYE—INTROMISSION THEORIES

There was an almost equally old but different view of vision among the Greek Natural philosophers, namely that vision involves something entering the eye from the object seen, a class of visual theory known as intromission theory. The first intromission theories were those of the atomists such as Democritus (ca. 410 BCE) and Epicurus (ca. 341–270 BCE). They believed that isomorphic images (or *eidola*) streamed off objects and entered the eye, where they were sensed.[7] As Epicurus put it,

> For particles are continually streaming off from the surface of bodies though no diminution of the bodies is observed...And those given off maintain their position and arrangement...it is by the entrance of something coming from external objects that we see shapes and think of them.[8]

The later atomist poet Lucretius (ca. 95–55 BCE) had a similar view. He called the images coming from objects *simulacra*, and in his poem *On the Nature of Things* compared them to the skin cast off by cicadas and snakes and the membrane (caul) covering the head of a newborn calf.[9]

In Aristotle (384–322 BCE) we find the first detailed discussion of vision. He argued that the atomist view is wrong because if objects put out copies of themselves, these would be objects themselves; but this is impossible because the copies would overlap on their way to the eye and two objects cannot be in the same place at the same time. The Alcmaeon-Plato extramission view is also inadequate because,

> In general it is unreasonable to suppose that seeing occurs by something issuing from the eye; that the ray of vision reaches as far as the stars, or it goes to a certain point and there coalesces with the object as some [Plato] think.[10]

Instead, Aristotle developed a rather complicated intromission theory. He assumed that a transparent medium, something like the modern ether, is found in air and water and is necessary for vision. Light is the state of this transparent medium. The color of an object (black and white are types of colors) moves the transparent medium and since the medium is continuous between the object and the eye, movement of the medium is sensed by the eye, yielding visual sensation.[11]

ALHAZEN'S SYNTHESIS

In Europe, soon after the deaths of Ptolemy and Galen, interest in studying the natural world declined and then virtually disappeared. Scientific inquiry gradually shifted to Islamic centers of learning, first in Baghdad and then in Cairo and Córdoba. Translation of Greek scientific works into Arabic began in the eighth century, and by the end of the ninth century, the achievements of Greek science were being actively discussed and often extended.[12]

The nature of vision and light were of great interest to Islamic scientists. Some natural philosophers such as Al-Kindi (ca. 801–866) defended and expanded Euclid's extramission views. Others such as Avicenna (Ibn Sina, 980–1037), probably the most important Arab natural philosopher, mounted an assault on extramission and built on Aristotle's theories of vision.[13] The primary achievement of Islamic visual science was to merge the two strains of Greek visual theory and eliminate the inadequacies of each. The architect of this synthesis was Ibn al-Haytham (965–1040), known in the West as Alhazen.[14] When translated into Latin in the beginning of the thirteenth century Alhazen's *Book of Optics* (*De Aspectibus*) dominated physiological optics in Europe for the 200 years until Kepler (1571–1630).

Alhazen's achievement had two parts. The first was to destroy extramission theory forever (at least among optical scientists) with a series of irrefutable arguments. For example, he pointed out that bright light produces pain in the eye and that when we look at the heavens it would hardly be possible for the eye to put out enough material to fill the space up to the stars. The second and more original contribution was to introduce a fundamentally new type of intromission theory that incorporated Euclid's rays and the visual cone of Ptolemy's extramission theory. Alhazen argued that although every point on a visible object sends light in every direction, only one ray from each point falls on the eye perpendicularly. All the others fall obliquely, are refracted and thereby weakened to virtual ineffectiveness. The sensitive part of the eye (the crystalline humor or lens, following Galen) responds only to the perpendicular rays and these form a cone with the visual field as the base and the center of the eye as the vertex (figure 3.1).[15]

Thus Alhazen not only eliminated extramission theory but also built a new intromission theory using the geometric ideas of Euclid and Ptolemy and the anatomico-physiologic ideas of Galen. His theory became "enormously influential" and became the basis of most of the subsequent work in optics in Europe between the thirteenth and seventeenth centuries. Indeed it led directly to Kepler's theory of the retinal image (1611) and modern visual science.[16]

The Fire in the Eye Is Quenched

Deformation phosphenes, the "fire in the eye" caused by pressure to the eyeball, continued to be observed after Alcmaeon and to demand explanation. Aristotle, having rejected the idea of light emitted from the eye, decided that phosphenes were due to "self-reflection" within the eye. Much later, Kepler still believed that pressure on the eye produced light. Since he realized that the retina was the sensitive surface he assumed that deformation of the eyeball produced sparks which stimulated the retina. He decided that the sparks were produced by mechanical irritation of the iris because "light can not possibly have its seat in the lens or vitreous body because then it would disturb the process of vision."[17]

Unlike Kepler, Descartes (1596–1650) rejected the idea of a physical light in the eye. Rather he suggested a blow on the eye produced vision in the same way he thought that light did, namely by moving the small fibers of the optic nerve. Newton (1642–1727) also thought that pressure on the eye, rather than producing light, mimicked the action of light on the retina:

> Do not these colors arise from such motions, excited in the bottom of the eye by the pressure and motion of the finger, as, at other times are excited there by light for causing vision?

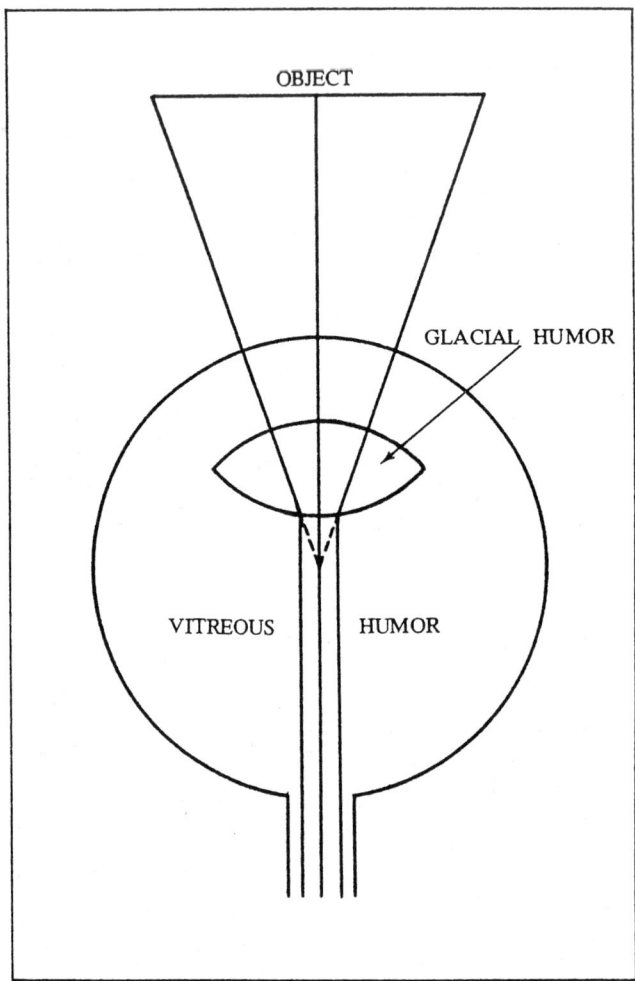

Figure 3.1
Alhazen's intromission theory of vision, which combines elements of earlier intromission and extramission theories. Only the rays from the object that fall perpendicular to the surface of the crystalline humor (Galen's term for our "lens") are sensed (Lindberg, 1992).

These and other speculations were offered as possible alternatives to fire in the eye as explanations for the phenomenon of phosphenes. However, the first experimental refutation of fire or light in the eye came in 1719 from the Italian anatomist Giovanni Morgagni (1682–1771). His experiment was very simple indeed. He pushed his eye to produce phosphenes and had his assistant look into his (Morgagni's) to see whether any light came out. He found:

> Even when [the assistant] observed extremely carefully and very bright light appeared to me [Morgagni] he could never observe any light by himself.

Georg Langguth (1711–1782), professor of anatomy and botany at the University of Wittenberg, extended Morgagni's observations. To find out whether light is generated in the eye he pushed his eye in the dark and, with a mirror, tried to see if light came out of his eye. Then, he wrote,

> A friend, who became curious about these phenomena . . . visited me in the dark room. I briefly explained to him what I was doing. The doors were closed and I asked him to observe my eyes very closely. While I was perceiving the small light [the phosphenes], he was not able to observe any small flashes or oscillating light. Thereafter, he performed the same experiment on himself . . . I could never discover any light leaving his eyes.

Thereafter, Morgagni and Langguth's experimental disproof of light in the eye was generally accepted, although their names gradually dropped out of the textbooks. For a modern view of the neural bases of deformation phosphenes see box 3.1. For some drawings of deformation phosphenes see figure 3.2.

Box 3.1
Neural Mechanism of Phosphene Formation

If phosphenes are not caused by a "fire in the eye," what are they caused by? Otto Grusser and his colleagues (1989) in Berlin studied the effect of eyeball deformation in the cat on the activity of retinal ganglion cells. They found that deformation caused a marked increase in the activity of retinal ganglion on-cells and a marked decrease in the activity of off-cells. Such a pattern of activity is certainly consistent with a phosphene-like perceptual effect. Grusser and colleagues suggested that the deformation caused retinal stretch, which in turn caused an increase in the surface of horizontal cells, which, he suggests, should depolarize them. Horizontal cell depolarization should indeed cause an increase in excitation of the on-ganglion cells and an increase in the inhibition of off-ganglion cells, the result they observed.

Persistence of Extramission Views

In spite of the decline of extramission theories under the widespread influence of Alhazen's *De Aspectibus*, and their disappearance among visual scientists after Kepler's demonstration of the inversion of the retinal image, extramission views remained and are still widely held. Extramission views may be found in at least four main arenas. The first is the widespread belief in the "evil eye." The second is in a long tradition in love poetry. Third and most surprisingly, strong extramission beliefs have been demonstrated in a high proportion of schoolchildren and college students in the United States. Finally, most people believe they can feel someone staring at them.

The Evil Eye

The evil eye approached and the storm sent no rain ... the milk was no longer plentiful ... the vigor of men was restrained. (Sumerian incantation, ca. 4000 BCE)

A glance of the Medusa turned men to stone.

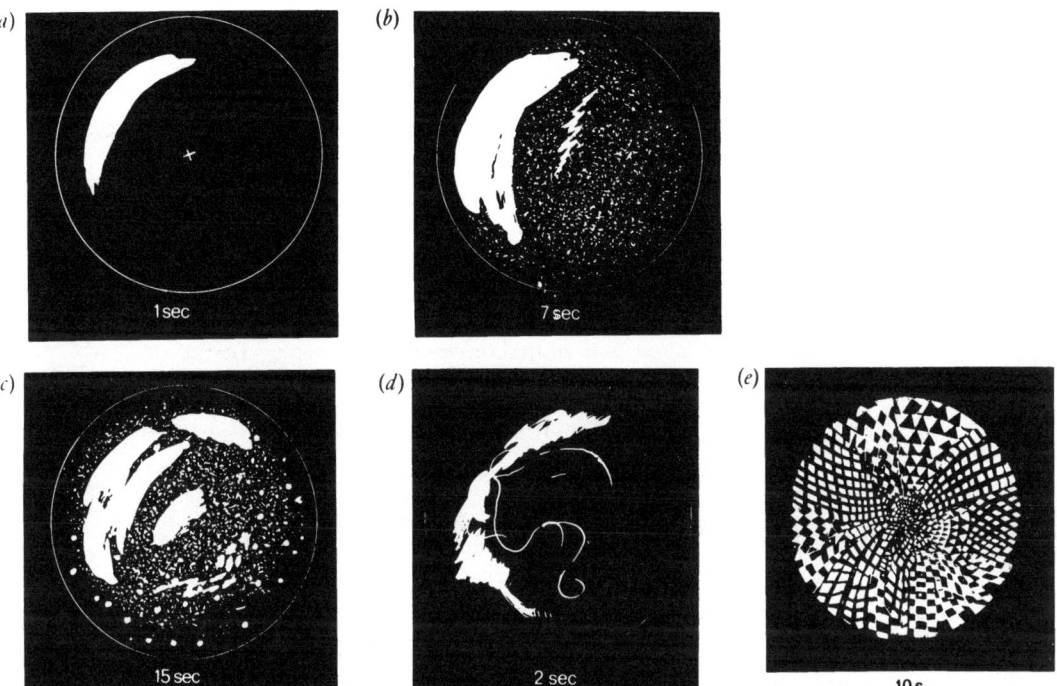

Figure 3.2
(a–c) Development of deformation phosphenes after pressure on the right temporal eyeball at different intervals and (d) after release. (e) simultaneous bilateral indentation of both temporal eyeballs produces a patterned and flickering phosphene (Grusser and Hagner, 1990). Reprinted with kind permission from Springer Science + Business Media.

Simon ben Johai and Rabbi Jochanan could with their looks transform people into a heap of stones. (Talmud)

Almost would the infidels strike thee down with their very looks when they hear the warning of the Koran. (Koran)

Witches may kill by their looks. (G. Mackenzie, *Laws and Customs of Scotland*, 1674)

A survey in 1962 at American University Hospital in Beirut indicated that 81% of 379 Armenian, Maronite Christian, and Sunni Muslim mothers sampled believed the evil eye affected their infant's health.

The foregoing are all examples of the "evil eye," the belief that there are individuals whose glance can produce harm, disease, or death.[18] Fear of the evil eye may be one of the oldest and most widespread superstitions. Freud called it "the most uncanny and universal." It is ubiquitous among cultures of Semitic and Indo-European origin and those that have come under their influence. The evil eye is usually the envious eye, and thus is often directed against the innocence of babies, the beauty of women, and the wealth of the powerful and is often attributed to the outcast, the ugly, and the other.[19]

There are a virtually infinite variety of preventives against the evil eye: spitting, gestures, charms, incantations, and amulets that vary from community to community. Some appear to be very old, such as making the sexual gesture, "fig" or "fico," by putting the thumb between the first and second finger, which is reported to be of Roman origin and is still common in older Italian and Jewish communities in New York. Again among older, more traditional people in this country, when a child or valued object is praised, the praise is often coupled with such phrases as "God bless it" among Irish and Italians and "keinahora" (no evil eye) among Jews.[20] (See figures 3.3 and 3.4.)

Figure 3.3
Drawings of evil eyes. Illumination accompanying a prayer against the evil eye on an Ethiopic scroll in the collection of Princeton University (Isaac, 1980). From The Princeton Collection of Ethiopic Manuscripts, Princeton University, Princeton, New Jersey.

Figure 3.4
Amulet against the evil eye (Courtesy E. Isaac). The Hebrew inscription exhorts the evil
eye to keep away. Sometimes a representation of an eye is found instead of a central text.
Among Jews this configuration is known as a hamesh hand or hand of Miriam, and
among Arabs as a hamsa hand or hand of Fatima. It is found both in this bilaterally symmetric form and in a more realistic one with only one thumb. Similar amulets and wall
plaques are readily found in the Middle East and in "New Age" shops around the world.

There have been a number of different interpretations of the resiliency and power of the superstition of the evil eye, ranging from the psychoanalytic to comparison with gaze aversion in primates.[21] What is clear is that the evil eye is the most widespread example of belief in something coming out of the eye, a very powerful extramission belief indeed.

Love Beams

My lady carries love within her eyes;
All that she looks on is made pleasanter,

Whatever her sweet eyes are turned upon,
Spirits of love do issue thence in flames

In such eyes as hers are
One surely stands whose glance can murder men

For me, out of her eyes comes the sweet light
That makes me heedless of each other lady;

These quotations from four poems of Dante Alighieri are in a tradition of love poetry extending from the classical poets through Arabic poetry to the Renaissance and beyond.[22] In this tradition the eyes of the Lady shoot arrows, darts, or fiery beams to induce love in the beholder, a tradition that has been termed the "the aggressive eye topos."[23] This theme seems to derive from Plato, as in the passage about his extramission theory of vision quoted above from the *Timaeus* and in his discussion of love in the *Phaedrus*.

There often seems to be a close affinity between the evil eye and the love arrows that the eye sends in the poetry of courtly love.[24] Indeed the third of the above quotes from Dante may be an example of this. On the other hand the eye beams in Donne's *The Extasie* seem more innocent and mutual than deadly or envious:

Our eye-beams twisted and did thred
Our eyes, upon one double string;[25]

Other quotes illustrating beams of love issuing from the eye are given in box 3.2.

Extramission among Schoolchildren and College Students

Piaget observed that children seem to think that seeing involves something coming out of the eye and even noted the similarity of this view to pre-Socratic extramission theory.[26] Inspired by Piaget's observation, Gerald Winer and Jane Cottrell carried out an extensive and systematic examination of the views of children and adults about the nature of vision and particularly whether it involves something going out of the eye or something entering the eye.[27]

When they asked whether something goes out of the eyes in the process of seeing, 57% of elementary school children and 33% of college students said yes. When asked to choose among "in," "out," or "both" as answers, 75% to 80% of the children and 24% to 33% of the college students gave one of the two extramission answers ("out" or "both"). Furthermore, among those who choose extramission about 90% of the schoolchildren and 77% of the college students thought the eye's output aided vision and 59% to 63% thought it was necessary. Winer and Cottrell found essentially the same level of belief of extramission under a great variety of different conditions and ways of asking the question, for example whether the questions or answers were verbal or pictorial, oral or written, and whether they were about luminous or nonluminous objects.[28]

Winer and Cottrell found that from the third to the eighth grade, the belief in extramission tended to decline and the belief in intromission tended to increase, a change that was more pronounced in the college students. However, the incidence of college students believing in extramission was little changed "as a function of having received lessons, reading and tests on perception in introductory psychology classes" or "having received readings on visual perception, immediately prior to [the] tests."[29]

There has been considerable research on "naive physics" indicating that children and adults often have erroneous beliefs about such things as

Box 3.2
Love and Extramission

The flaming rays of your lightning-like eye,
Instantaneously pierce my heart

—*Olivier de Magny* (Donaldson-Evans, 1980)

For your eyes, lady, caught and held me fast

—*Francesco Petrarca* (Lind, 1954)

The sparkling Glance that shoots Desire,
Drench'd in these waves, does lose its fire.

—*Andrew Marvell* (Gardner, 1957)

What joyes shall seize thy soul, when she
Bending her blessed eyes on thee
(Those second smiles of Heav'n) shall dart
Her mild rayes through thy melting Heart

—*Richard Crashaw* (Gardner, 1957)

Love-darting eyes...

—*John Milton* (Bartlett, 1956)

Then flash'd the living lightning from her eyes,
And screams of horror rend th' affrighted skies.

—*Alexander Pope* (Bartlett, 1956)

If beams from happy human eyes
Have moved me not;

—*Robert Lewis Stevenson* (Bartlett, 1956)

Lesbia hath a beaming eye
But no one knows for whom it beameth.

—*Thomas Moore* (Oxford Dictionary)

Lo! as that youth's eyes burned at thine, so went
Thy spell through him, and left his straight neck bent

—*Dante Gabriel Rossetti* (Oxford Dictionary)

A lover's eye will gaze an eagle blind

—*William Shakespeare* (Oxford Dictionary)

trajectories of falling objects.[30] However, there is little in ordinary experience that would contradict these "naive" or "intuitive" views nor are the correct views normally taught in elementary school. By contrast, anti-extramission experience such as the discomfort from looking at a bright light is common, and the elements of vision such as the inversion of the image on the retina are repeatedly taught in school. As Winer and colleagues put it after more than twenty studies on the subject, "the source and apparent strength of extramission beliefs in children and adults is somewhat of a mystery."[31]

The Feeling of Being Stared At

In 1898 the distinguished professor of psychology E. B. Titchener wrote in *Science*:

> Every year I find a certain proportion of students, in my junior classes, who are firmly persuaded that they can "feel" that they are being stared at from behind, and that a smaller proportion believe that they have the power of making a person seated in front of them turn around and look them in the face.[32]

After much discussion of this feeling (after all, he was the great champion of Introspection psychology) Titchener concluded

> I have tested . . . the "feeling of being stared at," at various times, in a series of laboratory experiments conducted with persons who declared themselves either peculiarly susceptible to the stare or peculiarly capable of "making people turn around." As regards such capacity, the experiments have invariably given a negative result.[33]

A later study followed this up and found that 68% to 86% of the students in a college class claimed to have the feeling of being stared at.[34] Since this "feeling of being stared at" implied some sort of belief in something coming out of the eye, that is, an extramission view, Cottrell and Winer included questions about staring in some of their studies of extramission described above.

Confirming the earlier studies, they found that 93% of college students said they could "feel the stare of other people."[35] Surprisingly, the proportion giving this answer went up with grade level, so that the percentages for the first, third, and fifth grades were 68%, 75%, and 80% respectively. This belief in feeling stares was clearly different from some kind of belief in extrasensory perception since "thinking about a person was not necessary to having one's gaze felt by another."

The finding that the ontogenetic trend for belief in the ability to feel stares was opposite to that for the belief that there are emissions from the eyes implies that the two extramission beliefs are somewhat different. Apparently, the belief in the efficacy of staring is more developmentally advanced than the belief that vision involves something leaving the eye.

Cognitive Development and the History of Science

There are several cases of striking similarities between the beliefs of children and naive adults and the theories held by premodern scientists. One example concerns motion. Naive or intuitive ideas about the motion of inanimate objects (such as the path of an object dropped by a moving person or of an object emerging from a curved tube) very closely resemble the "impetus" theory held by fourteenth-century Aristotelians.[36] Another example is the relationship between heat and temperature. Very similar ideas about the identity of heat and temperature are held by naive moderns as were held by a group of seventeenth-century Italian scientists forming the Accademia del Cimento and known as "the Experimenters."[37]

Is the belief in extramission among children and many naive adults another parallel between the ontogenesis of cognition and the history of science? Certainly there are similarities between Greek extramission theory and naive beliefs about vision. However, the parallels between the stages of ontogenetic development and historical development of visual science may be somewhat less compelling than for motion and heat. Before Alhazen, intromission theories, however incorrect, were held over the same time period as were extramission theories. Furthermore, at least one type of extramission theory, that of belief in the detectability and efficacy of staring, increases rather than decreases with general cognitive development.

————

POSTSCRIPT

The decade since this article was written has been a good one for Ibn al-Haytham (known in the West as Alhazen). The first three books ("On direct vision") of his pioneering *Kitab al-Manazir* or *Optics* were finally translated into English with commentary.[38] A version of this work had been available as a Latin manuscript by the end of the twelfth century, and was printed in 1572 as *De Aspectibus*. The first five books (of seven) of the Latin translation were recently translated into English, making the most important book on optics and visual science until the seventeenth century now readily available with detailed glosses.[39] Ibn al-Haytham was given a glowing portrait in *Science*, which stressed his work in mathematics as well as optics.[40] A hagiographic account on Wikipedia claims he was "the originator of experimental science and experimental physics," and the "founder of experimental psychology." The article ranks *Optics* with Newton's *Principia*.[41] Ibn al-Haytham has craters on the moon named after him, and his portrait is on the current Iraqi ten-thousand-dinar note (figure 3.5) and in a *New Yorker* cartoon of that note.

Ibn al-Haytham and David Hockney

Ibn al-Haytham has become involved in a major controversy about the development of realistic painting in Europe. The contretemps began when the British (and California) painter David Hockney suggested that the sudden

————

Figure 3.5
An Iraqi ten-thousand-dinar note showing Alhazen. The diagram is from Alhazen's copy of an Arabic translation of the Greek mathematician Apollonius's *Conics*. Apparently, Alhazen earned his living by copying translations of Greek mathematics books (Sabra, 1983).

arrival of photorealistic painting in Europe around 1420 (as in Van Eyck's *Arnolfini Wedding* and *Cardinal Albergati*) was due to the use of optical devices. The physicist Charles Falco came to his support and suggested that a concave mirror and later a convex lens had been the optical aids used. Falco further suggested that the use of these optical instruments came from Alhazen's *De Aspectibus*.[42]

However, the consensus of historians of science, historians of art, and optical scientists is that the material and textual evidence and the detailed examination of the relevant paintings do not support the Hockney-Falco thesis.[43] Furthermore, A. M. Smith, Alhazen's primary student today, insists that there is nothing in Alhazen or the Pespectivist tradition that he founded that indicates any understanding of the use of mirrors or lenses to project images.[44] Alhazen did build and experiment with a camera obscura (pinhole camera), and he and his Pespectivist followers were very much interested in what it revealed about the properties of light (as was Leonardo). However, it would be of no use in painting until a convex lens was put in the aperture and there is no evidence for that before 1550.[45] In any case, there is no question that artists used optical projection devices as aids to their painting later, from the middle of the seventeenth century, such as Vermeer (1632–1675) and Vanvitelli (1652–1736).[46] There does seem to have been a lack of interest by art historians in the use of optical and mechanical aids in painting, but this may be reduced by Hockney's provocative idea even if his specific proposal does not appear to be supported by the evidence.

Undergraduates and Extramission

Winer and Cottrell have continued and expanded their extraordinary demonstration that children, undergraduates, and adults believe, as did Alcmaeon and Plato, that vision involves something coming out of the eye. They showed that this extramission view persists among undergraduates even directly after reading and classroom instruction about vision. In fact, explicit instruction with diagrams and arrows that vision involves intromission and not extramission had only a transient effect.[47] They speculate that the strength of extramission views may be related to the "outer-orientated" quality of seeing.

NOTES

This article, which first appeared in *The Neuroscientist* (5: 58–64 [1999], "The Fire that Comes from the Eye") started as a review of Greek visual theory, but then Muslim optics, the evil eye, and undergraduate intuitions were added. The postscript extends the subject to a current controversy about early Dutch painting.

1. Gross, 1998a.

2. Theophrastus, 1917.

3. Gross, 1998a.

4. Plato, 1959.

5. Cohen and Drabkin, 1958.

6. Lindberg, 1976.

7. Theophrastus, 1917; Lindberg, 1976.

8. Lindberg, 1976.

9. Lindberg, 1976.

10. Lindberg, 1976.

11. Lindberg, 1976.

12. Lindberg, 1992.

13. Lindberg, 1976.

14. Gross, 1981.

15. Lindberg, 1976, 1992.

16. Lindberg, 1976.

17. The quotations in this subsection are taken from Grusser and Hagner, 1990.

18. All the evil-eye quotes are from Dundes, 1981.

19. Dundes, 1981.

20. Dundes, 1981; Gifford, 1958. The other day my young Princeton dentist said "keina-hora" to me, explaining that it was his Jewish mother's expression against bad luck.

21. Dundes, 1981.

22. Lind, 1954; Cline, 1972; Donaldson-Evans, 1980.

23. Donaldson-Evans, 1980.

24. Cline, 1972; Spence, 1996.

25. Gardner, 1957.

26. Piaget, 1979.

27. Cottrell and Winer, 1994; Winer and Cottrell, 1996; Winer et al., 1996a; Winer et al., 1996b.

28. Cottrell and Winer, 1994; Winer and Cottrell, 1996; Winer et al., 1996a, Winer et al., 1996b.

29. Cottrell and Winer, 1994; Winer and Cottrell, 1996; Winer et al., 1996a; Winer et al., 1996b; Winer and Cottrell, 2004.

30. E.g., McCloskey and Kargon, 1988.

31. Winer et al., 1996a.

32. Titchener, 1898.

33. Titchener, 1898.

34. Coover, 1913.

35. Cottrell, Winer, and Smith, 1996.

36. McCloskey and Kargon, 1988.

37. Wiser and Carey, 1983.

38. Alhazen, 1989.

39. Alhazen, 2001, 2006. The sources for the translation from the Arabic and for the Latin translation and the relationships between the two are discussed in Alhazen, 1989.

40. Rashed, 2002.

41. http://en.wikipedia.org/wiki/Ibn al-Haytham, accessed February 28, 2007.

42. Hockney, 2001; For citations to a series of joint papers by Hockney and Falco as well as Falco's recent reply to his critics, see Falco, 2007a. Ibn al-Haytham's contribution is proposed in Falco 2007b.

43. A Web site presenting many sides of the debate on the Hockney-Falco thesis on the use of optical imagery in early Renaissance painting is http://webexhibits.org/ hockneyoptics/post/intro.html, accessed February 19, 2008, which includes essays by Hockney and Falco, their chief critic David Stork, and many others. Stork's more recent arguments are Stork and Duarte, 2007, and papers cited therein. The evaluations of historians of science, which are largely negative on the Hockney-Falco thesis and are not well represented on this Web site, may be found in a special issue of the journal *Early Science and Medicine* (Dupré, 2005).

44. Smith, 2005. Smith spent many years translating and explaining *De Aspectibus*.

45. Kemp, 1990.

46. Dupré, 2005.

47. Gregg et al., 2001; Winer et al., 2002; Winer et al., 2003; Winer and Cottrell, 2004.

4

The Discovery of Motor Cortex

The modern neurophysiology of the cerebral cortex began in 1870 with the discovery by Gustav Fritsch (1838–1927) and Edmund Hitzig (1838–1907) that electrical stimulation of the cerebral cortex produces movements. Their discovery was important for several reasons. First, it was the first clear experimental demonstration of a region of the cerebral cortex involved in motor function. Second, it was the first evidence that the cortex was electrically excitable. Third, it was the first experimental evidence of a topographically organized representation of the body in the brain. Finally, it was the first strong experimental evidence for localization of function in the cerebral cortex. Overall, as Fritsch and Hitzig somewhat immodestly put it,

> by the results of our own investigations, the premises for many conclusions about the basic properties of the brain are changed not a little.... some psychological functions, and perhaps all of them ... need circumscribed centers of the cerebral cortex.[1]

This chapter discusses their experiment and its background in the previous two centuries. Fritsch and Hitzig's basic findings were soon replicated by David Ferrier (1843–1928).[2] We consider the differences between the two studies in both method and interpretation and how these differences have continued to reverberate in research on motor cortex.

FRITSCH AND HITZIG'S EXPERIMENT

When Fritsch and Hitzig carried out their famous experiments they were young medical Privatdozents (roughly, assistant professors) at the Physiological Institute in Berlin. Previously Fritsch had worked on electric fish and carried out anthropological and geographical studies and served as a battlefield surgeon in the Franco-Prussian war.[3] Hitzig was a psychiatrist who had tried the therapeutic use of electricity on his patients. After their collaboration Hitzig continued to work as a psychiatrist and to research on cortical localization in animals.[4] Fritsch returned to his earlier interests in anthropology and was particularly concerned with using studies of the eye and hair to establish the superiority of the white race. In the course of the former he coined the terms "fovea" and "area centralis."[5] When the American anatomist C. L. Herrick (founder of the *Journal of Comparative Neurology*) met Fritsch and Hitzig after they were famous he described them as "splendid specimens of physical development and German culture at its best."[6] Hitzig, who came from a distinguished assimilated Jewish family, was characterized by one of his biographers as "a stern and forbidding character of incorrigible conceit and vanity complicated by Prussianism."[7] Portraits of Fritsch and Hitzig are shown in figure 4.1.

In their now classic experiment, Fritsch and Hitzig strapped their dogs down on Frau Hitzig's dressing table, as there were no animal facilities at the Physiological Institute.[8] In their early experiments they used no anesthesia or analgesic, although ether surgical anesthesia had been introduced in 1846 and morphine analgesia in 1803.[9] Later they did use "morphine narcosis." They began by removing the cranium and cutting the dura, the dog showing "vivid pain." They stimulated the cortex with platinum wires with "galvanic stimulation": brief pulses of monophasic direct current from a battery at the minimum current that evoked a sensation on their tongue.

Figure 4.1
Gustav Fritsch (left) and Eduard Hitzig.

The usual response to this stimulation was a muscle twitch or spasm (*Zuckung*). Their central findings were that (a) the stimulation evoked contralateral movements, (b) only stimulation of the anterior cortex elicited movements, (c) stimulation of specific parts of the cortex consistently produced the activation of specific muscles, and (d) the excitable sites formed a repeatable, if rather sparse, map of movements of the body laid out on the cortical surface (figure 4.2). They went on to show that lesion of a particular site impaired the movements produced by stimulation of that site. The loss of function was not complete, suggesting to them that there were other motor centers that had not been impaired by the lesion.[10]

THE SITUATION BEFORE FRITSCH AND HITZIG

The background of Fritsch and Hitzig's discoveries lay in earlier developments stretching back many centuries.

Before the Eighteenth Century

From the earliest Western medical writings it was thought that the movement of the body was controlled by the brain. In the Edwin Smith Surgical Papyrus, whose origins lie in the Pyramid Age (about 30th century BCE) there are a number of descriptions of motor dysfunctions after head injury.[11] For example, in case 5 the patient "walks shuffling with the sole on the side of him having that injury which is to his skull" (presumably a contracoup injury where a blow to one side of the head causes the brain to impact on the inside of the contralateral skull).

The Hippocratic doctors (5th century BCE) wrote extensively on the treatment of head wounds and, unlike the author of the Surgical Papyrus, were well aware that head injuries produce contralateral symptoms. However, they were primarily interested in diagnosis and treatment and had little interest in studying the underlying anatomy or physiology.[12]

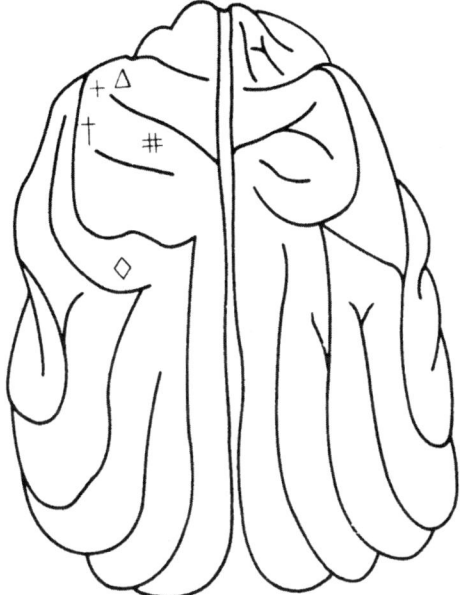

Δ: Stimulation produced twitching of neck mucles.
+ : Abduction of foreleg.
†: Flexion of foreleg.
#: Movement of rear leg.
◊ : Facial twitching.

Figure 4.2
Drawing after Fritsch and Hitzig's (1870) figure of stimulation sites on the dog's cortex.
Note that the topography is not impressive (Brazier, 1988).

Galen (129– ca. 213) was the most important figure in classical medicine and biology and a brilliant experimental physiologist and anatomist.
(See chapter 2.) His ideas dominated European medicine for more than
1,500 years. This was especially true for his views of brain function and the
control of movement. His theories derived from many sources including
his training in Alexandria (where human vivisection had been practiced[13]),
his clinical experience (especially as physician to a gladiator school) and his
experiments on the spinal cord and brain of animals. Galen distinguished
between sensory and motor nerves; he thought nerves were hollow and carried "psychic pneuma." The brain was supposed to act as a pump that
moved the psychic pneuma from the sense organs into the ventricles, then
into the motor nerves and finally into the muscles, causing their contraction
by inflation.[14]

René Descartes (1596–1650) elaborated these ideas by suggesting that
the centrally located pineal body directed the pneumatic flow from sense
organs into the muscles to expand them.[15] Several lines of evidence soon
refuted this pneumatic theory of movement. Francis Glisson (1597–1677)
and, independently, Jan Swammerdamm (1637–1680) demonstrated that
contraction of a muscle did not increase its volume, as it should if the
pneuma were swelling it.[16] Alexander Monroe (1697–1762) showed that
ligating a nerve produced no distal swelling and nothing flowed from a cut
nerve. Having disproved pneuma as a transmitter of nerve activity to muscle, Monroe suggested that, instead, electricity might be the mechanism.[17]

Electricity and the Nervous System

The eighteenth century was a period of great activity and interest in the
new discoveries about electricity in both the salons and laboratories of the
time. Among the intriguing gadgets were electrostatic machines, the Leydon
jar (the original capacitor) and the gold leaf electroscope (which detects
static electricity). At this time it was realized that man-made electricity and
lightning were the same phenomenon as that found in the electric fish, an

animal whose shocking properties had been known since classical times. There were a number of attempts to use electricity for therapeutic purposes, including by the French revolutionary Jean-Paul Marat (1743–1793) and the American savant and revolutionary Benjamin Franklin (1706–1790) as well as some ineffectual studies of electrical stimulation of various brains from frog to dead human.[18]

The modern study of the electrical nature of nervous activity began with Luigi Galvani's (1737–1798) demonstration that electrical stimulation of the sciatic nerve in a severed frog's leg resulted in contraction of the attached muscle. These findings sparked a fierce debate between Galvani and Alessandro Volta (1745–1827) concerning the source of the electricity that caused the legs to become "reanimated." Volta considered the cause to be the use of dissimilar metals, whereas Galvani was convinced that the electricity came from within ("animal electricity").[19] Eventually this led, on one hand, to the development of the electric battery and, on the other hand, through the work of Emil du Bois-Reymond (1818–1896), Julius Bernstein (1839–1917) and others to the discovery of the action potential. Galvani's results soon prompted attempts to stimulate other nervous structures, but the vast majority of experiments on the electrical stimulation of the cerebral cortex were negative, reinforcing the prevailing view that the cortex had no significant functions.[20]

Cortex as "Rind"

In the eighteenth century, the cerebral cortex was usually dismissed as an insignificant "rind," which indeed the Latin "cortex" means. The first to microscopically examine the cortex was Marcello Malpighi (1628–1694), professor in Bologna, the founder of microscopic anatomy and discoverer of capillaries. He saw it as made up of little glands with attached ducts. Similar "globules" were reported by many subsequent microscopists.[21] Perhaps they were observing pyramidal cells. At least in Malpighi's case, artifact is a more likely possibility, since modern studies have shown that his globules are more prominent in boiled tissue (which is what he used) than in fresh tissue.[22] Malpighi's view of the brain as a glandular organ was commonly held in the seventeenth and eighteenth centuries, perhaps because it fit with the much earlier, but still persisting, Aristotelian view that the brain was a cooling organ and the Hippocratic theory that it was the source of phlegm.[23]

Another common eighteenth-century view of the cortex was that it was largely made up of blood vessels. One of the earliest advocates of this idea was Frederik Ruysch (1638–1731), professor of anatomy in Amsterdam, who noted: "the cortical substance of the cerebrum is not glandular, as many anatomists have described it, nay have positively asserted, but wholly vascular."[24] In this view the convolutions were viewed as mechanisms for protecting the delicate blood vessels of the cortex.

Willis Gives the Cortex Cognitive Function

Prior to the nineteenth century there were only a very few figures who advocated significant functions for the cerebral cortex. The first and most important of these was Thomas Willis (1621–1675), who held the Chair of Natural Philosophy at Oxford and was one of the founders of the Royal Society. His *Anatomy of the Brain* was the first monograph on the brain and dealt with brain physiology, brain chemistry, and clinical neurology as well as brain anatomy.[25] Many of its illustrations (such as figure 4.3) were by the architect Sir Christopher Wren, then professor of astronomy at Oxford.

Willis implicated the "cortical and grey part of the cerebrum" in the functions of memory and will. In his scheme, sensory signals came along the sensory pathways into the corpus striatum where the common sense was located. They were then elaborated into perceptions and imagination in the overlying white matter (then called the corpus callosum, or hard body, since it was harder than the cortex) and then passed to the cerebral cortex where they were stored as memories. According to Willis, the cortex

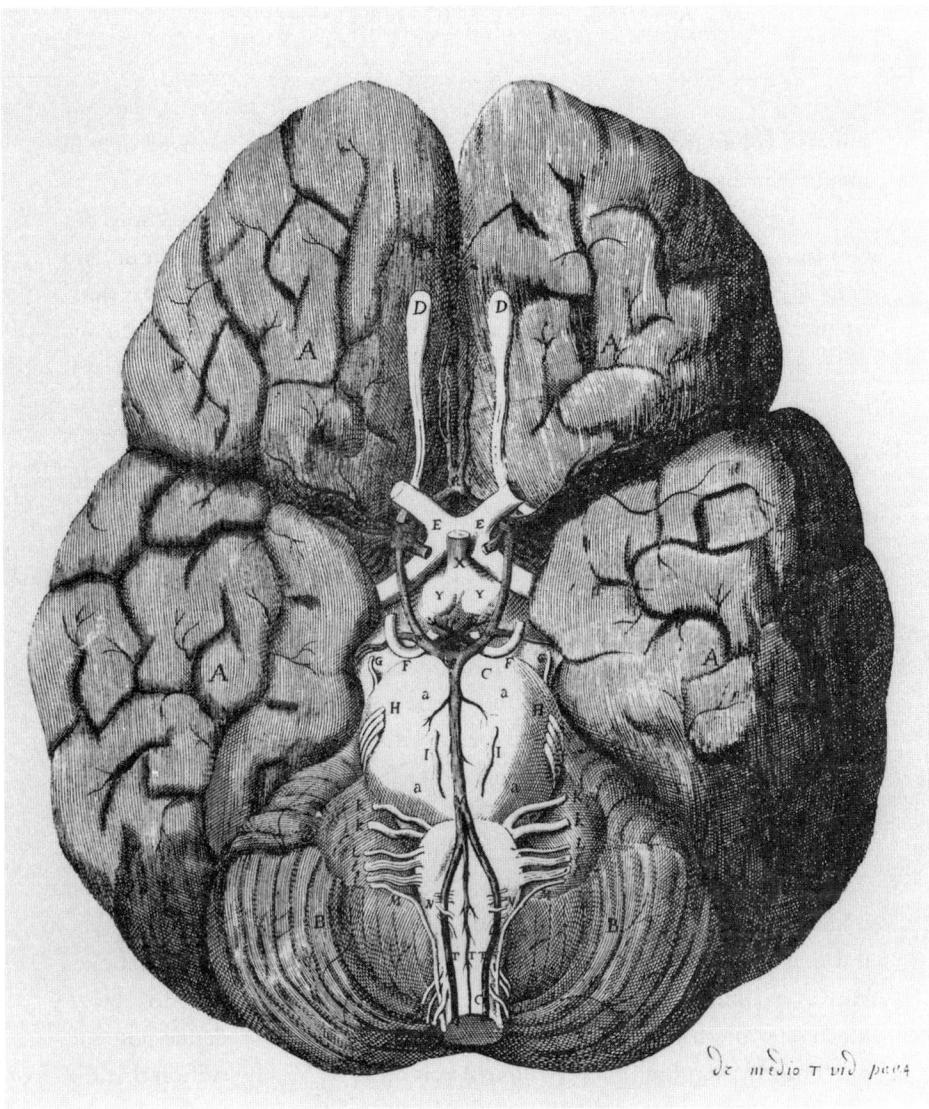

Figure 4.3
Ventral view of the brain from Willis, *Cerebri Anatome* (1664), drawn by the architect Sir Christopher Wren. Note the detailed drawing and labeling of the cranial nerves and basal brain structures (including the circle of Willis) in contrast to the vague and partially obscured representation of the cerebral cortex, all of which has the single designation *A*.

initiated voluntary movement whereas the cerebellum was involved only in involuntary movement.

Willis's ideas on brain function came not only from his dissections but also from his experiments on animals and his correlation of symptoms and pathology in humans. Willis noticed that whereas the cerebellum was similar in a variety of different mammals, the complexity of the cerebral convolutions varied greatly among different animals and this variation was correlated with intellectual ability.

In spite of the relative importance of the cerebral cortex in Willis's schema, there is no detailed drawing of the cortex in his works: he apparently never asked Wren or anybody else to produce one (see figure 4.3). In fact, for another 150 years the cortex continued to be drawn as Erasistratus of Alexandria (200 BCE) had first suggested: as coils of the small intestine (figure 4.4).[26]

Although Willis was a major figure in his time, his ideas on the importance of the cerebral cortex soon disappeared and the views of the cortex as a glandular, vascular, or protective rind returned to their original dominance. There were two figures who did challenge this view. The first was François Pourfour du Petit (1664–1741), a French army surgeon. He carried out a series of systematic experiments on the effects of cortical lesions in dogs and related them to his clinicopathological observations on wounded soldiers.[27] From these studies he realized that the cerebral cortex plays a critical role in normal movement and that this influence is a contralateral one. However, his observations were totally ignored until they were rediscovered much later. Perhaps this was because he did not hold an academic post and published his account in a very limited edition. Yet, his observation that the cortex was insensitive to touch was repeatedly cited to support the views of von Haller, who, as discussed below, was the dominant physiologist of the day. Thus, du Petit's work demonstrating motor functions of cortex was probably ignored largely because of the anticortex ideology of the time, not because it was published in a minor journal.

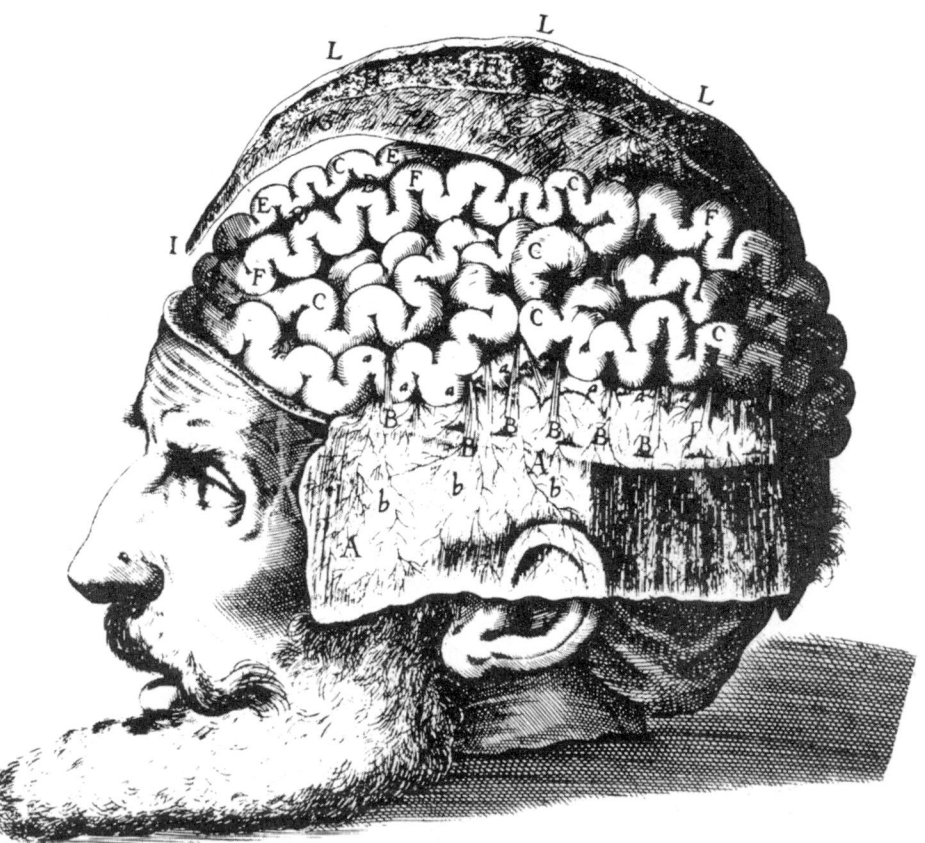

Figure 4.4
The depiction of the cerebral convolutions by Giulio Casserio (1561–1617). The convolutions are not differentiated in any way and, following Erasistratus, look like intestines (Clarke and Dewhurst, 1972).

Swedenborg's Lost Speculations on Cortex

The second major eighteenth-century figure advocating the importance of the cortex was Emanuel Swedenborg (1688–1772), the founder and mystical prophet of the "New Jerusalem" or Swedenborgian Church (which is still active in the United States and Great Britain). On the basis of reviewing the literature, Swedenborg arrived at an amazing set of prescient ideas on the importance of the cerebral cortex in sensation, cognition, and movement.[28]

He argued that the cortex was the highest sensory and motor structure of the brain. In an anticipation of neuron theory, he called the cortical "glandules" described by Malpighi "little brains" (*cerebella*) and suggested that they were functionally independent and connected by thin fibers (see figure 1.11 in my previous *Tales*[29]). These fibers also ran through the white matter and medulla down to the spinal cord and then by way of nerves to parts of the body. The operations of these cerebella were the basis of sensation, mentation, and movement. He seems to have had the idea of somatotopic organization of motor function in the cerebral cortex: He (correctly) localized control of the foot in the dorsal cortex (he called it the "highest lobe"), the trunk in an intermediate site, and the face and hand in the ventral cortex, his "third lobe."[30]

Swedenborg's philosophical and religious writing had a major impact on European and American philosophers, writers, and artists. Indeed, he continues to be a subject of religious, philosophical, and fringe science tracts. However, he never had any impact on the scientific study of the brain. There is no evidence that contemporary physiologists and anatomists even read his writings on the brain. He never held an academic post or had students, colleagues, or even scientific correspondents. He never seems to have carried out any systematic empirical work on the brain, and his speculations were in the form of baroque and grandiose pronouncements embedded in lengthy books on the human soul by one whose fame was soon to be that of a mystic or madman. Furthermore, some of his more advanced ideas, such as on the organization of motor cortex or the functions of the pituitary gland, did not appear in print until after they were no longer new.[31]

Haller Declares the Cortex Insensitive

In spite of Willis, du Petit, and Swedenborg, who all thought the cortex was a crucial brain structure, the opposite view was very much the dominant one in the eighteenth century. Much more representative and influential was Albrecht von Haller (1708–1777), professor at Tuebingen and later Bern, who dominated physiology in the middle of the eighteenth century.[32] In his monumental *Elementa Physiologiae Corporis Humani* and his *Icones Anatomicae*, he divided the organs of the body, as well as parts of the nervous system, into those "irritable" (such as muscle) and those "sensible" (such as the sense organs and nerves). Using animals, he tested the "sensibility" of various brain structures with mechanical stimuli such as picking with a scalpel, puncturing with a needle, and pinching with forceps as well as with electrical and chemical stimuli, such as silver nitrate, sulfuric acid, and alcohol. With these methods he found the cortex completely insensitive. By contrast, he reported the white matter and subcortical structures to be highly sensitive; their stimulation, he said, produced expressions of pain and movement. He concluded that all parts of the cortex had the same function because they were equally insensitive and all subcortical regions were also equivalent because their stimulation had equal effects. Because of his prestige and many students and followers, Haller's views of the insensitivity and equipotentiality of cortex apparently suppressed contrary observations and ideas on the importance of cortex and persisted well into the next century.[33]

Phrenology Calls Attention to the Cortex

The systematic study of the localization of different psychological functions in different regions of the cerebral cortex begins with Franz Joseph Gall (1758–1828) and his collaborator J. C. Spurzheim (1776–1832), the founders of phrenology. Gall and Spurzheim viewed the brain as an elabo-rately wired machine for producing behavior, thought, and emotion and the cerebral cortex as a set of organs for carrying out these functions. Their phrenological system was an attempt to relate psychological functions to the organs of the cerebral cortex—to relate brain and behavior. Phrenology was based on four assumptions: (1) Intellectual abilities and personality traits are differentially developed in each individual. (2) These abilities and traits reflect faculties that are localized in specific organs of the cerebral cortex. (3) The development or prominence of these faculties is a function of the activity and therefore the size of the cortical organs. (4) The size of each cortical organ is reflected in the prominence of the overlying skull, i.e., in cranial bumps.[34] Phrenological localizations of psychological functions are shown in figures 4.5 and 4.6. The drawing in figure 4.5 is arguably one of the first accurate drawings of the human cortex, reflecting the importance Gall and Spurzheim gave to the cerebral cortex.

These otherwise reasonable hypotheses had one fatal flaw: the nature of the evidence. Gall and Spurzheim relied almost entirely on obtaining supportive or confirmatory evidence. They collected large numbers of skulls of people whose traits and abilities were known, examined the heads of distinguished savants and inhabitants of mental hospitals and prisons, and studied portraits of the high and low on various intellectual and affective dimensions. Throughout, they were seeking confirmation of their initial hypothesis, which usually derived from a few cases. For example, the idea for a language organ in the frontal lobes come from Gall's experience of a classmate who had a prodigious verbal memory and protruding eyes (being pushed out by a well-developed frontal lobe, Gall thought). The idea for an organ of destructiveness (or carnivorous instinct) came from examining the skulls of a parricide and of a murderer that were sent to him, from noticing its prominence in a fellow medical student who "was so fond of torturing animals that he became a surgeon," and from examining the head of a meat-loving dog he owned. All their methods were used to seek confirma-tions; contradictions were explained away.[35]

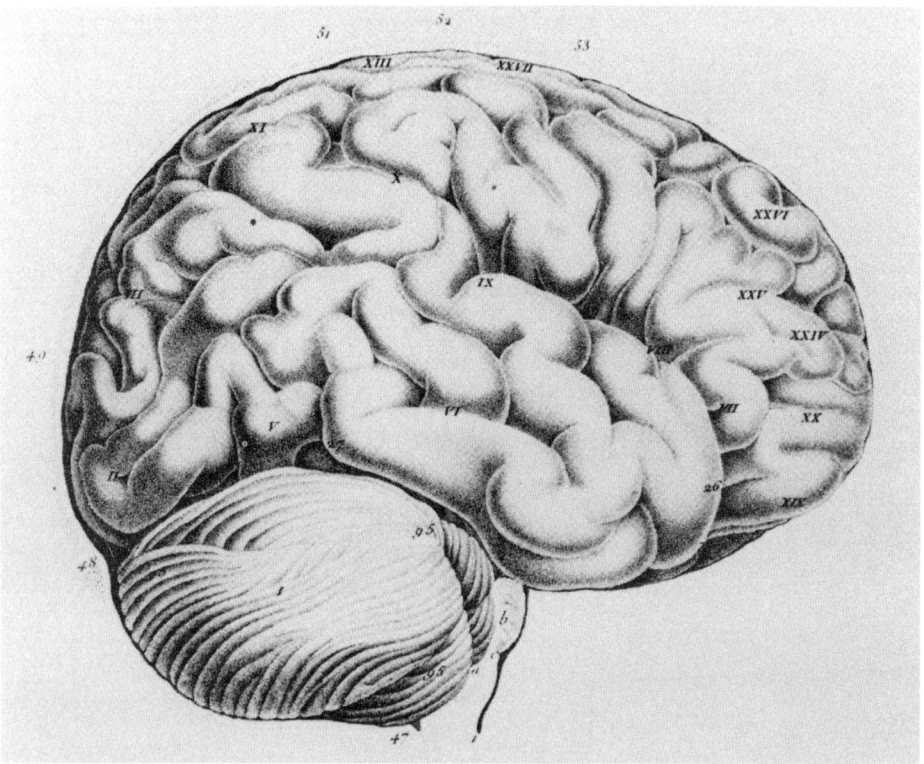

Figure 4.5
This accurate and aesthetic drawing of the cerebral cortex is from Gall and Spurzheim, 1810–1819, vol. 1. The Roman numbers refer to the organ's qualities as follows. Organs common to men and animals: I, instinct of reproduction; II, love of offspring; III, friendship; IV, self-defense and courage; V, carnivorous instinct, tendency to murder; VI, cunning, cleverness; VII, ownership, covetousness, tendency to steal, love of authority; VIII, pride, arrogance, haughtiness; IX, vanity, ambition, love of glory; X, caution, forethought; XI, memory of things and facts, educability; XII, sense of places and spaces; XIII, memory and sense of people; XIV, memory of words; XV, sense of language and speech; XVI, sense of color; XVII, sense of sound, music; XVIII, sense of numbers, mathematics; XIX, sense of mechanics. Organs in man alone: XX, wisdom; XXI, sense of metaphysics; XXII, satire, witticism; XXIII, poetical talent; XXIV, kindness, compassion, morality; XXV, mimicry; XXVI, religion; XXVII, firmness of purpose, obstinacy, constancy.

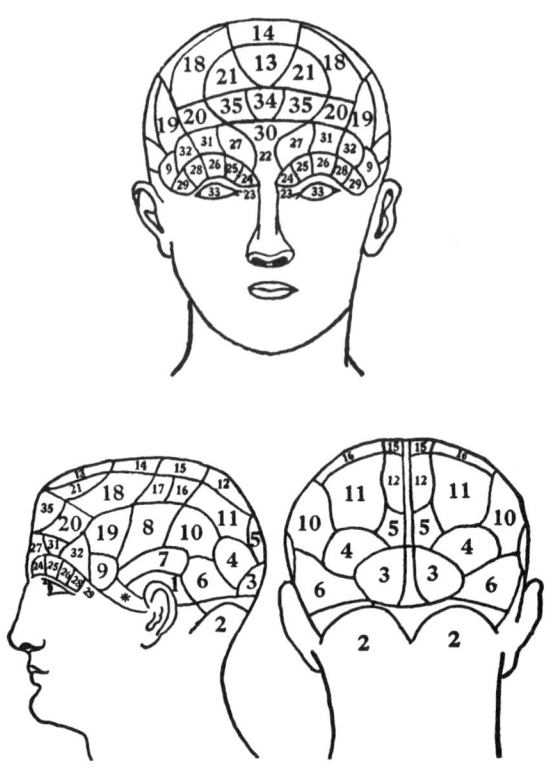

AFFECTIVE FACULTIES

PROPENSITIES	SENTIMENTS
? Desire to live	10 Cautiousness
* Alimentiveness	11 Approbativeness
1 Destructiveness	12 Self-Esteem
2 Amativeness	13 Benevolence
3 Philoprogenitiveness	14 Reverence
4 Adhesiveness	15 Firmness
5 Inhabitiveness	16 Conscientiousness
6 Combativeness	17 Hope
7 Secretiveness	18 Marvelousness
8 Acquisitiveness	19 Ideality
9 Constructiveness	20 Mirthfulness
	21 Imitation

INTELLECTUAL FACULTIES

PERCEPTIVE	REFLECTIVE
22 Individuality	34 Comparison
23 Configuration	35 Causality
24 Size	
25 Weight and Resistance	
26 Coloring	
27 Locality	
28 Order	
29 Calculation	
30 Eventuality	
31 Time	
32 Tune	
33 Language	

Figure 4.6
Phrenological head is from Spurzheim, 1834. In comparison to figure 4.5 note that the locations of the faculties are more specific, as are their names.

Gall and Spurzheim gave cogent reasons why the use of experimental lesions in animals (or "mutilations" as they called them) or the study of brain-injured humans was inadequate, in their view, to test their ideas. The state of anatomical knowledge was insufficient to remove a single organ. The organs were interconnected so that removal of one could alter the function of another. Animals would often live only a few days after surgery with the available methods. Generalizing from animals to humans was hazardous. Previous studies had not looked at complex cognitive or emotional function in animals but only such things as "sensibility" or irritability. Needless to say, however, when a clinical case that supported their view was brought to them, they would cite it as further confirmatory evidence.[36]

Gall and Spurzheim's cortical localizations were of "higher" intellectual and personality traits. They accepted the prevailing view that the highest sensory functions were in the thalamus and the highest motor functions in the corpus striatum. The term "phrenology" was never used or approved by Gall. Originally he used the term "craniology" then "organology" and finally "functions of the brain" or "brain physiology" to describe his theory of punctate localization of psychological function in the brain. In 1813, Spurzheim separated from Gall and moved to England, where he substituted the term "phrenology." There were a number of differences between Gall's organology and Spurzheim's phrenology, which continued to develop after they separated. Gall was much more tentative in both naming functions, often giving several terms and long description and a more approximate location whereas Spurzheim was satisfied with a single term. Gall had 27 organs (19 of which were found in animals) whereas Spurzheim extended them to 32 to 37, with sharper boundaries; later phrenologists such as Combe added more (compare figures 4.5 and 4.6). Gall continued to emphasize empirical observation whereas Spurzheim became increasingly speculative and used phrenology as a way of predicting future behavior. Furthermore, Spurzheim was much more interested in applying the phrenology to social reform and to the treatment of mental disease. The term "phrenology" was generally applied to both sets of ideas.[37]

Phrenology met with considerable opposition from political and religious authorities, particularly on the Continent, largely because it was viewed as implying materialism and determinism and denying the unity of the mind (and soul) and the existence of free will. On the other hand, phrenology spread widely, particularly in the United States and Great Britain, both as a medical doctrine and as a "pop" psychology. As a medical doctrine it influenced the diagnosis and treatment of mental and neurological disease. It generated widespread interest both among the general populace and among such writers and savants as Honoré de Balzac, Charles Baudelaire, George Eliot, August Comte, Horace Mann, Alfred Russell Wallace, and George Henry Lewis. In fact it rapidly became a popular fad and drawing-room amusement, particularly in Great Britain and the United States. Phrenological societies and journals continued to flourish in both countries well into the twentieth century.[38]

Flourens Attacks Gall

In the scientific world the most important and influential critique of Gall came from Pierre Flourens (1794–1867). Starting in the 1820s and continuing for over twenty years Flourens carried out a series of experiments on the behavioral effects of brain lesions, particularly with pigeons. Flourens reported that lesions of the cerebrum had devastating effects on willing, judging, remembering, and perceiving. However, he found that the site of the lesion was irrelevant: all regions of the cerebrum contributed to these functions. The only exception was vision, in that a unilateral lesion produced only contralateral blindness, but again there was no localization within the hemisphere. These holistic results tended to eclipse Gall's ideas of punctate localization, but only in scientific circles and only temporarily.[39]

Flourens's finding of cognitive losses after cerebral lesions was actually a confirmation of Gall's emphasis on the cognitive role of the cortex, a viewpoint that had been virtually absent before Gall. The cortex had been termed a "superadded" structure lying hierarchically and physically above the highest sensory structure, the thalamus, and the highest motor structure, the corpus striatum. This view of the "higher" functions of the cortex combines Haller's view of the insensitivity of cortex and both Gall's and Flourens's attribution of "higher faculties" to the cortex.[40]

Broca "Confirms" Gall

In spite of the bitter attacks of Flourens, Gall's idea of punctate localization and even many of his specific localizations such as language in the frontal lobes and sexuality in the cerebellum continued to be actively debated in the middle of the nineteenth century. At least in the scientific community, the supposed correlations between skull and brain morphology were quickly recognized as erroneous. Yet, Gall's ideas stimulated the search for correlations between the site of brain injury and specific psychological deficits in patients as well as in experimental animals. Reports of such correlations were published in both the phrenological and mainstream neurological literature, and the question of the localization of psychological function in the brain was hotly debated at scientific meetings.[41]

Thus, in 1848, J. B. Bouillard (1796–1881), a powerful figure in the medical establishment and a supporter of Gall, offered a cash prize for a patient with major frontal lobe damage who did *not* have a language deficit. The debate about localization reached a climax at a series of meetings of the Paris Société d'Anthropologie in 1861. At the April meeting, Paul Broca (1824–1880), a founder of the society, announced that he had a critical case on this issue. A patient with long-standing language difficulties, nicknamed "Tan" because that was all that he could say, had just died. The next day Broca displayed his brain at the meeting and indeed it had widespread damage in the left frontal lobe (actually much more widespread than today's "Broca's area"). He presented several similar cases in the next few months. Not only did Broca's cases finally establish the principle of discrete localization of psychological function in the brain, but the discovery itself was hailed as a vindication of Gall. Broca himself had regarded Gall's work as "the starting point for every discovery in cerebral physiology in our century."[42]

In spite of its absurdities and excesses, phrenology facilitated the development of the study of the brain and behavior in several ways: by stressing that the human mind could be subdivided into specific functions and that specific brain mechanisms underlay specific mental abilities and behavior; by emphasizing the importance of the cerebral cortex; and by stimulating a great surge of research on the psychological effects of human brain damage, experimental lesions in animals, and attempts to electrically stimulate the brain. After Gall, less radical divisions of brain function, such as those of Flourens, Fritsch and Hitzig, and Ferrier, were much more readily accepted. The search for organs in the cerebral cortex led directly to cytoarchitectonics and myeloarchitectonics—attempts to distinguish cortical areas on the basis of structure—and to the tracing of sensory pathways to specific cortical areas. The phrenologists also influenced the development of physical anthropology: they were the first to develop methods and instruments for measuring the cranium.[43]

The cytoarchitectonic, PET, functional MRI, and other imaging maps of the cerebral cortex that are now ubiquitous in neuroanatomy, neurophysiology, and neuropsychology textbooks bear more than a coincidental resemblance to phrenological charts. They are the direct descendants of the ambitious, albeit heavily flawed, program of phrenology to relate brain structure and behavior.

John Hughlings Jackson

The first intimations of a somatotopically cerebral motor mechanism (aside from Swedenborg's lost ideas) came from John Hughlings Jackson (1835–1911), who is often called "the father of English neurology" because of his many neurological discoveries. Jackson reasoned that the cerebrum should have basic sensory-motor functions and gathered clinicopathological evidence for this view. In studying epileptic seizures, including those of his wife, he noticed that there was a consistent systematic spread of convulsions from one body part to the next. From this he inferred that different parts of the brain must be involved in the control of different muscle groups and that these parts must be arranged in a way to mimic the organization of the body.[44] Before 1870, he appeared unsure whether this somatotopy was found in the corpus striatum, the traditional highest motor structure, or in the cortex. But by 1870, Jackson placed the somatotopically organized structures that are responsible for movement primarily in the convolutions.[45]

Fritsch and Hitzig had not mentioned Hughlings Jackson in their first paper, an omission that upset David Ferrier, an admirer of Jackson who replicated Fritsch and Hitzig's 1870 study as described below[46]. However, it should be noted that Jackson did not clearly implicate cortex until 1870, and then only in the *Transactions of St. Andrews Medical Graduates' Association*, which the Germans may well have not seen.[47] Much later Hitzig commented that his experiments with Fritsch "had only confirmed the conclusions reached by Jackson on clinical evidence."[48] Jackson himself later claimed he had been clearer about the role of cortex earlier than his numerous, difficult and ambiguous writings in scattered medical journals demonstrate.[49]

The Experiments on Motor Cortex

What led Fritsch and Hitzig to electrically stimulate the cortex of a dog? As described above, this was a period of very active exploration of the nervous system with electrical stimulation. While there were reports of effects of stimulation of the spinal cord and brain stem, all attempts at eliciting effects of stimulation of the cerebral cortex had been universally ineffective, including those made by such skilled experimenters as Albrecht von Haller, François Magendie (1783–1855), and Carlo Mattecci (1811–1868). As Fritsch and Hitzig put it in the introduction to their paper,

---

Even in other fields than in physiology, there can hardly be a question about an opinion which seems to be so unanimous, which seems to be so completely settled as that of the excitability of the cerebral hemisphere.[50]

One impetus to their experiment was the paradox that some central nervous system structures were excitable and yet the cortex didn't seem to be. Another was their own previous observations. Hitzig had tried electrical stimulation of the human head for therapeutic purposes and had noticed it caused eye movements.[51] He then tried rabbits and also elicited movements. Fritsch, while working as a battlefield surgeon, is said to have noticed that the contralateral limbs twitched while dressing an open head wound.[52]

Ferrier's Replication and Subsequent Developments

In 1866, James Crichton-Browne (1840–1938, later Sir) was appointed head of the West Riding Lunatic Asylum in West Riding of Yorkshire.[53] In addition to transforming the West Riding Lunatic Asylum into an enlightened mental hospital, he made it a major center for research on brain anatomy, brain physiology, and brain pathology with laboratories, weekly seminars, visiting lecturers, and visiting researchers. He also founded the first journal devoted to brain research, the *West Riding Lunatic Asylum Medical Reports*, which published a series of important papers for seven years.[54]

Soon after the Fritsch and Hitzig paper appeared, Crichton-Browne invited his fellow medical student and Scotsman David Ferrier to come to the West Riding hospital and research center to follow up the Germans' work on motor cortex.[55] Ferrier had been heavily influenced by John Hughlings Jackson and realized that Fritsch and Hitzig had confirmed Jackson's ideas. He was eager to study the relation of their findings to Jackson's observations about epilepsy as well as to determine the generality of their findings across a variety of species, especially monkeys.

During and after his work on motor cortex, discussed in the next two sections, Ferrier studied the effects of stimulating throughout the accessible cortex and cerebellum. In some of these experiments, Ferrier thought that he might be stimulating sensory rather than motor areas. For example, when he stimulated the superior temporal lobe he elicited ear movements, which he interpreted as suggesting some auditory functions, and stimulating the parietal lobe resulted in eye movements, suggesting some visual functions. He then set out in a long series of experiments to test these possibilities by studying the behavioral effects of lesions of frontal, temporal, parietal, and occipital cortex. These experiments, described in papers, two editions of *The Functions of the Brain* and *The Localization of Cerebral Disease*, were major contributions to the understanding of the cerebral cortex.[56] They also resulted in Ferrier being charged under the Cruelty to Animals Act in 1881 (see box 4.1) and being knighted in 1912. Ferrier continued his clinical work and is said to have been one of the last physicians to conduct rounds in "the traditional top hat and black tailcoat."[57] A portrait of Ferrier is shown in figure 4.7.

Ferrier's Experiments on Motor Cortex

Ferrier's initial experiments on motor cortex were carried out at West Riding primarily on dogs but also on cats, rabbits, jackals, and other animals.[58] He asked three main questions: could seizures similar to those observed by Hughlings Jackson be elicited by electrical stimulation? What was the effect of stimulation beyond the limited region explored by Fritsch and Hitzig? Were the effects of stimulation similar in different species?

Ferrier used faradic stimulation, a type of alternating current (see figure 4.8) and he usually stimulated for longer durations than Fritsch and Hitzig. He found he could induce seizures in the variety of animals tested as long as the duration of the electrical stimulus was 5 seconds or more. The seizures, he thought, were strikingly similar to the "marching seizures" observed by Hughlings Jackson in his wife and other patients.

Box 4.1
Ferrier's Arrest

In England in the 1870s there was an intense campaign to ban all experiments on
live animals ("vivisection"; French, 1975). Private bills were introduced into Par-
liament, European physiologists who practiced vivisection like François Magendie
and Claude Bernard (see chapter 8) were vilified, and Disraeli, then Prime Minis-
ter, was besieged by antivivisectionist memos from Queen Victoria. The most militant group was the Society for the Protection of Animals from Vivisection,
known as the Victoria Street Society because of the location of its headquarters.
Its leader was the very formidable Frances Power Cobbe (1822–1904). Many of
its members were abolitionists, feminists, vegetarians, and opponents of immuniza-
tion. The Victoria Street Society demanded total abolition of animal experimentation. Finally, in response to the widespread antivivisectionist agitation, a Royal
Commission was set up. (One of the very few occasions that Darwin left Down
House in Kent to come to London was to testify at the Royal Commission against
antivivisection legislation.) Its report led to the Cruelty to Animals Act of 1876
that regulated animal experimentation in the United Kingdom until 1986. This
act required annual licenses for specific experimental procedures at specific sites.
The required protocols had to justify the experiments in terms of medical applications and to minimize pain. It was a compromise that certainly did not satisfy the
more militant antivivisectionists. They were determined to prove it ineffective in
order to get it strengthened (French, 1975; Richards, 1987; Rupke, 1987; Elston,
1987; Hampson, 1987).

   They got their opportunity in 1881 at the International Medical Congress
in London. This was the site of a dramatic and widely reported debate between
Ferrier and Friedrich Goltz (1834–1902), the leading opponent of localization of
function in the cerebral cortex. To defend his arguments, Ferrier displayed a monkey with partial paralysis after a motor cortex lesion (the French neurologist J.-M.
Charcot was amazed and, in French, exclaimed "it's a patient") and one with a superior temporal gyrus lesion (which was shown to be deaf when it did not flinch
when a pistol was discharged near its head). By contrast, Goltz demonstrated that
his dogs had their motor and sensory functions relatively intact even after supposedly massive cortical lesions. Ferrier claimed this was because Goltz's lesions were
incomplete. To resolve the issue, the dogs and monkeys were immediately sacrificed and their brains turned over to a panel of experts. The panel found that
Goltz's dogs had much smaller lesions than he had claimed (sparing what we now
know to be primary sensory and motor cortex). Ferrier was vindicated (Anonymous, 1881b; Klein et al., 1883).

Box 4.1
(continued)

The Victoria Street Society had been closely following what scientists were writing and saying and matching it against the published list of those licensed under the act to carry out experiments. They immediately saw that Ferrier did not have a license for the surgery on the monkeys he described at the Congress and had him prosecuted under the act. The defense successfully argued that Ferrier was not liable because his collaborator G. F. Yeo, not Ferrier himself, had actually carried out the surgery and Yeo did have a proper license. The outcome of the trial was viewed as a major defeat by the antivivisectionists. However, the medical and scientific communities were frightened by the fact of the trial and began to organize to protect themselves (Anonymous, 1881b; French, 1975).

At least up to 1975 there had been only two other prosecutions under the act.

Ferrier explored the effects of stimulation of much more of the cortical mantle than Fritsch and Hitzig had, and indeed much more than is considered motor cortex today (see figure 4.9). Like Fritsch and Hitzig, Ferrier found that he could evoke movements of different body parts from different regions of cortex. He localized centers related to the movements of the eyelids, face, mouth, tongue, ear, neck, hand, foot, and tail. Furthermore, Ferrier realized how similar the organization of these "centers" were to the map hypothesized by Hughlings Jackson on the basis of his study of human seizures.

In comparing dogs, cats, and rabbits, Ferrier noticed that the overall organization was the same in different species, but that some parts of the body were more represented in some species. He suggested that these differences depended "on the habits of the animals." This is the first suggestion of a relationship between the organization of the brain and behavioral specializations in a species.

In 1874, Ferrier moved to Kings College Hospital in London and expanded his work to monkeys. By doing so he hoped to understand the organization of the motor centers in the human brain. In fact, his localizations

Figure 4.7
David Ferrier.

Figure 4.8
Du Bois–Reymond inductorium used for faradic stimulation. Although there were many
different versions of the inductorium, it seems likely that Ferrier used this version, popular in the 1870s, in his stimulation experiments. The primary coil is used to induce a current in the secondary coil. The current was reversed using an electromagnet (b), which
opened and closed the circuit using a key (f). The strength of the stimulus was varied by
varying the distance between the primary and secondary coils. From Klein et al., 1873.

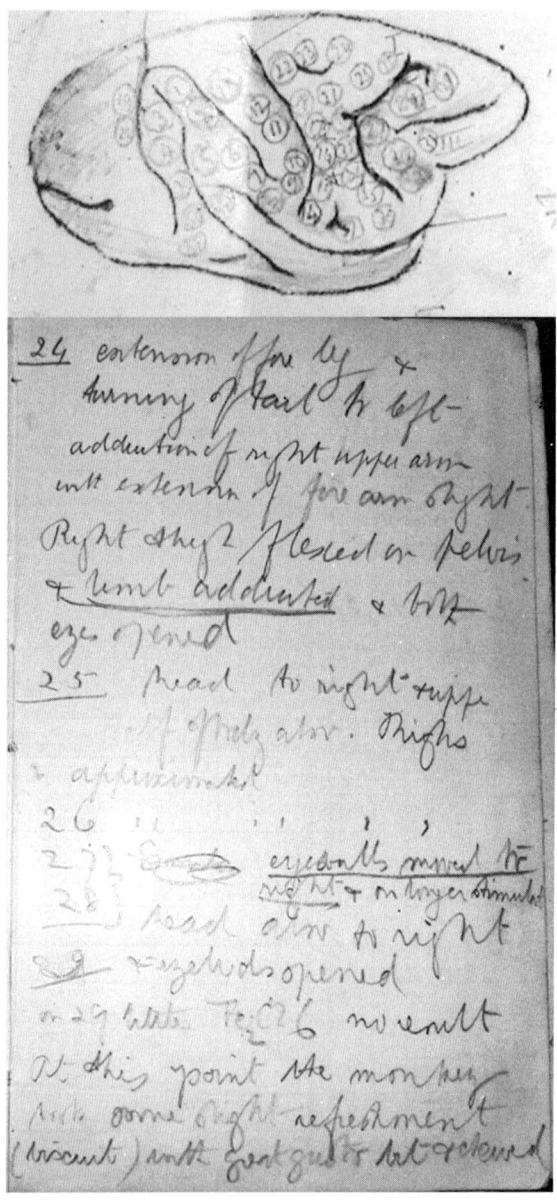

in monkeys were soon transferred to the human brain and used to localize and remove tumors and blood clots from the brains of patients, with surprising success and accuracy.[59] In the monkey brain he again found that movements of various body parts were localized in different regions of cortex. In monkeys, Ferrier delineated 19 centers related to different movements such as walking, arm retraction, flexion and extension of the wrist, mouth opening and "protrusion of the tongue," "sneering expressions of the face," and eye movements.[60] Figure 4.9 from his unpublished lab notebook shows examples of these.

As had Fritsch and Hitzig, Ferrier found that movements that appeared to be represented at a site would be impaired when he destroyed that site. Furthermore, the larger the lesion, the larger the affected part of the body: lesions of the entire motor area resulted in paralysis of one side of the body; lesions of "centres" that, when stimulated, caused movements of the hand and foot "caused motor paralysis of the same movements and of none other";[61] finally, small lesions of only the "centre for the biceps" caused "the right arm to hang by the right side in a state of flaccid extension... [the monkey] had lost the power of flexing the right arm."[62] As described

Figure 4.9
Pages from Ferrier's laboratory notebook recording the sites at which he applied stimulation (top) and the results of that stimulation (bottom). In the experiment recorded here Ferrier evoked complex movements of various parts of the body using cortical stimulation. The notebook reads:
24. extension of foreleg and turning of tail to left adduction of right upper arm with extension of forearm slight right thigh flexed on pelvis and limb adducted and both eyes opened
25. head to right and upper half of body also thighs approximated
26. Same as 25.
27 and 28. eyeballs moved to right and on longer stimulation hand also to right and eyelids opened
At this point the monkey took some slight refreshment (biscuit) with great gusto bit and chewed
From the collection of Ferrier's notebooks from 1873 to 1883 in the archives of the Royal College of Physicians, London (MS 246).

in box 4.2, Ferrier initially had some difficulty in getting his results published in the *Philosophical Transactions of the Royal Society* because the referees thought he had not given adequate credit to the prior work of Fritsch and Hitzig.

Differences between the Fritsch-Hitzig and the Ferrier Studies

The basic findings of Fritsch and Hitzig and of Ferrier were very similar. They both found that electrically stimulating the frontal cortex produced discrete contralateral movements. They both found that movements of different body parts could be evoked from different regions of cortex. They both found that when a part of the motor cortex was damaged, the animal appeared to be unable to execute the movement that is evoked by stimulation. However, there were some important differences in their results, their methods, and their interpretations that have influenced the subsequent direction of research on motor cortex.

The chief difference in their results was the type of movements that each evoked, indeed that each aimed to evoke. Fritsch and Hitzig described the movements they evoke in terms of spasms or twitches (*Zuckungen*) of a small number of adjacent muscles. By contrast Ferrier described the movements that he observed in terms of the natural movements they resembled, as in the following examples from different species. In the cat, Ferrier notes that stimulation of a particular site causes "the shoulder to be raised, and the limb to be adducted, exactly as when a cat strikes a ball with its paw." In the dog, stimulation of the supraorbital region caused the following sequence of movements: "The animal opens its mouth, retracts the upper lips, and makes a sort of sniffing or snarling noise." In the rabbit, during stimulation of the frontal region, "The mouth is drawn to the left, and a munching movement of the left side of the mouth is made, as if the animal is eating."[63] And likewise, in the monkey, stimulating of one site causes "flexion with outward rotation of the thigh, rotation inwards of the leg, with flexion of the toes—the action being such as is seen when the monkey makes a

Box 4.2
Ferrier Quarrels with Referees over Priority

Ferrier's first two papers on motor cortex involving dogs, cats, and other infrapri-
mate species (1873, 1874a) were published in his institute's house organ *The West
Riding Lunatic Asylum Medical Reports*. Then in 1874 he submitted a paper on stimulation of motor cortex in dogs, monkeys, and other animals to *The Philosophical
Transactions of the Royal Society*, then the most prestigious scientific journal in Britain if not the world. The paper began by acknowledging the influence of Hughlings Jackson and then mentioned that the findings confirm and extend the work
of Fritsch and Hitzig. Initially, the paper was sent to two distinguished referees,
Michael Foster (professor of physiology at Cambridge) and George Rolleston
(professor of anatomy at Oxford). Both criticized the paper for, among other
things, not giving adequate credit to Fritsch and Hitzig's prior work: it "failed to
do justice to them" in Foster's words.

The editor of the *Transactions* then called in a third referee, T. H. Huxley,
perhaps the most politically powerful scientist at the time, "for the purpose of
ascertaining... whether Dr. Ferrier has or has not done sufficient mention of the
labour of his predecessors in the same field of investigation." Huxley concluded
he had not. The views of the referees were sent to Ferrier, and he slightly revised
and resubmitted the paper. In that unpublished revision he wrote:

To Fritsch and Hitzig belong the honour of the discovery of the method of exciting the
functional activity of the brain, and of showing that in the anterior part of the brain
there are localized regions, the stimulation of which causes certain constant muscular
movements of the head and limbs. The exact extent to which they carried out their
localization experiments will be found under the head of Experiments in dogs page
53.... Since this paper was written Hitzig has published a new series of experiments
and a criticism of my paper in the West Riding Reports ["Untersuchungen über das
Gehirn," 1874]. Hitzig differs from me in several important points in regard to method
and extent of localization, and his criticisms deserve to be noticed. I have preferred
however to leave the present paper as it is, and to reserve a consideration of Hitzig's
researches as well as those of other experimenters in the localization of function in the
brain for a future occasion.

The revised paper was still unacceptable because it still gave inadequate credit
to Fritsch and Hitzig, since, for example, its "acknowledgement of Hitzig's claim
to priority was so slight" that its publication would be "unfortunate to English
Science."

Box 4.2
(continued)

> Later that year Ferrier presented the material in this paper at a meeting of
> the Royal Society and a three-page abstract (unrefereed) was published in its *Pro-*
> *ceedings* (Ferrier, 1874b). The abstract described the work as a confirmation of
> Hughlings Jackson's views and did not mention Fritsch and Hitzig. Ferrier then
> wrote up the monkey stimulation data in detail and it was accepted by the *Philo-*
> *sophical Proceedings of the Royal Society* (1874c). Although the paper did not even
> mention Fritsch and Hitzig, the priority issue must have seemed less critical since
> Fritsch and Hitzig had only worked on dogs and this paper was exclusively on
> monkeys. Much of the other material in the previously rejected paper was eventu-
> ally published in his 1876 book. Ferrier did give Fritsch and Hitzig more credit
> there, and by 1890 he wrote, "The whole aspect of cerebral physiology and pa-
> thology was revolutionized by the discovery, first made by Fritsch and Hitzig,
> that certain definite movements could be excited by the direct application of elec-
> trical stimulation to definite regions of the cortex cerebri in dogs" (Ferrier, 1890).
> (Ferrier's rejected paper, from which the other quotations in this box were taken,
> is in the archives of the Royal Society [AP.56.2]. The correspondence with Foster,
> Rolleston, and Huxley about Ferrier's rejected paper is in the archives of the
> Royal Society [RR.7.299-305; MC10.194. 247])

grasping movement, or scratches its chest or abdomen with its foot."[64] At
another site, Ferrier reported that stimulation caused the following movement: "Shoulder raised, forearm firmly flexed, hand clenched and supinated.
The hand ultimately raised to the mouth, the angle of which is retracted
and elevated." Ferrier, on the basis of replications in 12 other monkeys,
identified this site as "the centre for the biceps and muscles concerned in
bringing the hand to the mouth."[65]

In light of recent work on motor cortex there appear to be two main
reasons for the differences between the results of Fritsch and Hitzig and of
Ferrier, namely, the nature of the electrical stimulus and its duration.[66]

Fritsch and Hitzig used primarily galvanic stimulation whereas Ferrier
used primarily faradic stimulation, and both argued for the superiority of

their choice. Ferrier argued that faradic stimulation was critical for evoking complete movements from a motor center and therefore to "properly characterize" that center. Galvanic stimulation, he noted, results in "only a sudden contraction in certain groups of muscles, but fails to call forth the definite purposive combination of muscular contractions, which is the very essence of the reaction and the key to its interpretation."[67]

The second major difference between Fritsch and Hitzig and Ferrier was the duration and strength of stimulation. It was critical in determining whether the evoked movements were twitches or resembled purposive movements. Although Fritsch and Hitzig do not explicitly state the duration of stimulation used in their experiments, it is likely that they used stimulation durations that were very short—possibly less than a second—causing twitches or spasms that they describe as "fast-passing."[68] They were aware of the noxious effect that longer durations of galvanic stimulation could have on the cortex, and consequently advocated the use of short durations and weak currents. Furthermore, Hitzig noted:

> By the use of very weak currents these muscle contractions can be localized to specific narrowly limited muscle groups, with stronger currents, stimulation . . . led to the immediate participation of other muscles or even muscles of the corresponding body half.[69]

In contrast, Ferrier stressed the importance of longer durations of stimulation, in keeping with his emphasis on producing a complete movement, and consequently producing the participation of other muscles. In his experiments with different durations of stimulation he found:

> A slight stimulus of short duration causes only a part of a complex action, which is manifested in its completeness when the stimulus is of somewhat greater intensity and duration.[70]

In addition to differences in the nature of the movements that Fritsch and Hitzig and Ferrier evoked, Ferrier tested a much greater extent of the cortex. Ferrier located motor centers in several cortical regions, including eye movement areas now known as the frontal eye fields (area 8) and the lateral intraparietal area (LIP). Indeed he thought that posterior parietal cortex was the "visual center" because he elicited eye movements from stimulation there.[71] In later work Hitzig did investigate wider regions of cortex and described an "inexcitable zone" that overlaps with regions that Ferrier claimed to be excitable. Hitzig considered this to be due to Ferrier's general incompetence, such as using too strong a current and failing to replicate his observations sufficiently.[72] Ferrier thought Hitzig to have failed to produce movements from these regions because the currents he used were too weak.[73]

In summary, Fritsch and Hitzig obtained "twitches" of a few muscles from brief monopolar DC stimulation, whereas Ferrier obtained complex movements from long-duration AC stimulation.

COMMENT

Both Fritsch and Hitzig's and Ferrier's papers on motor cortex were greeted by much skepticism. Their results went against the generally accepted views that the striatum was the highest motor center, that the cortex was inexcitable, and that functional localization in the cortex was phrenological pseudoscience. The critics usually interpreted the evidence for the localization of motor function in cortex as artifactual, due to a "spread of current" to the striatum. To overcome these criticisms, Victor Horsley (1857–1916), C. S. Sherrington (1857–1952), and others began meticulous "punctate" mapping of cortex using the minimum current to elicit the smallest discernible movement.[74] Thus, Ferrier's search for the full movement became overshadowed by the effort to localize function more and more precisely.

In the century after the first electrical stimulation experiments on motor cortex, it became possible to evoke smaller and smaller movements, and therefore to localize motor functions in an even more "punctate" fashion. By the 1980s, it was possible to record small changes in the electromyographic (EMG) activity of muscles in response to single pulses of stimulation.[75] Leyton and Sherrington's map of motor cortex in the chimpanzee, followed by Penfield's human motor homunculus and Woolsey's maps of monkeys and other animals, became the standard picture of motor cortex.[76] One consequence of this effort to localize the effects of stimulation with increasing precision was that the emphasis of stimulation studies shifted to the representation of muscles, rather than of movements.[77] It was not clear, however, how these maps were related to movement. Did motor cortex control relatively low-level aspects of movement such as the muscle tensions required for the flexion and extension of individual joints? Or did it control more complex aspects of movement such as the coordinated muscle activity required for reaching?

The two contrasting views that first emerged with Fritsch and Hitzig's stress on brief muscle twitches and Ferrier's stress on integrated movements became central issues in the study of motor cortex. Some researchers emphasize the cortical control of small groups of muscles and individual joints;[78] others emphasize the cortical control of reaching in specific directions.[79] Recently, Graziano and colleagues have revisited Ferrier's idea that motor cortex may control complex, highly integrated behavior.[80]

Postscript

Graziano and colleagues electrically stimulated motor cortex with long durations, like Ferrier, and produced complex, coordinated, apparently natural movements, again reminiscent of those reported by Ferrier.[81] On the basis of these results they suggested that motor cortex is organized into regions that "emphasize different ethologically relevant categories of actions." They suggest that the organization of motor cortex may be understood as the result of the superimposition on a somatotopic map of other

---

"maps" such as of ethologically relevant movements and of hand position in space.[82]

Strick has criticized the electrical stimulation methodology underlying Graziano's results, but it is hard to see how they can be accounted for simply by artifactual spread of current.[83]

NOTES

This chapter derives primarily from a recent article in *The Journal of the History of the Neurosciences* (16: 320–331 [2007], "The Discovery of Motor Cortex and Its Background in the 18th and Early 19th Centuries") and from an earlier piece coauthored with Charlotte Taylor and published in *The Neuroscientist* (9: 332–342 [2003], "Twitches versus Movements: A Story of Motor Cortex"). The latter paper originated when Taylor, then a graduate student, and Michael Graziano, a postdoc, working in my lab on motor cortex, realized that the results they were finding were related to the nineteenth-century controversy on motor cortex between David Ferrier on the one hand and G. T. Fritsch and S. Hitzig on the other.

1. Fritsch and Hitzig, 1870.

2. Ferrier, 1873.

3. Grundfest, 1963.

4. Brazier, 1988; Hitzig, 1874.

5. Grundfest, 1963; Brazier, 1988; Fritsch, 1912.

6. Herrick, 1892.

7. Finger, 2000; Kuntz, 1953.

8. Kuntz, 1953.

9. Magner, 1992.

10. Fritsch and Hitzig, 1870.

11. Breasted, 1930.

12. Hippocrates, 1927.

13. Gross, 1998a.

14. Gross, 1998a, 1998b.

15. Descartes, 1972.

16. Fearing, 1970.

17. Brazier, 1959.

18. Finger, 2000; Brazier, 1959, 1984.

19. Finger, 2000; Brazier, 1984.

20. Finger, 2000; Brazier, 1959, 1984.

21. Meyer, 1971; Clarke and O'Malley, 1996.

22. Meyer, 1971; Clarke and Bearn, 1968.

23. Gross, 1995; Hippocrates, 1927.

24. Clarke and O'Malley, 1996.

25. Dewhurst, 1982; Willis, 1664.

26. Schiller, 1992.

27. Kruger, 1963.

28. Gross, 1997a.

29. Gross, 1998a.

30. Gross, 1997a.

31. Gross, 1997a.

32. Neuburger, 1981.

33. Neuburger, 1981.

34. Gall and Spurzheim, 1835; Young, 1970; Gross, 1999d.

35. Gall and Spurzheim, 1835; Young, 1970; Gross, 1999d; Spurzheim, 1834.

36. Gall and Spurzheim, 1835; Young, 1970; Gross, 1999d; Spurzheim, 1834.

37. Gross, 1999d; Spurzheim, 1834; Zola-Morgan, 1985.

38. Young, 1970; Cooter, 1985.

39. Gross, 1999d; Clarke and O'Malley, 1996.

40. Gross, 1999d.

41. Young, 1970; Gross, 1999d; Broca, 1960; Schiller, 1992.

42. Young, 1970; Broca, 1960; Schiller, 1992.

43. Young, 1970; Gross, 1999d.

44. Finger, 2000; Young, 1970; Temkin, 1971.

45. Young, 1970; Temkin, 1971; Jackson, 1870, 1958.

46. Ferrier, 1873.

47. Jackson, 1870.

48. Hitzig, 1900.

49. Jackson, 1875.

50. Fritsch and Hitzig, 1870.

51. Hitzig, 1870.

52. Walker, 1998.

53. The county of Yorkshire is divided into three ridings, or thirds, a North, East, and West riding.

54. Viets, 1938. Crichton-Browne then abandoned the *West Riding Lunatic Asylum Medical Reports* to found, with Hughlings Jackson and Ferrier, the journal *Brain*.

55. Ferrier, 1876.

56. Ferrier, 1873, 1875, 1876, 1878, 1886; Finger, 1994, 2000; Gross, 1998a.

---

57. Clarke, 1970.

58. Ferrier, 1873, 1874a.

59. Anonymous, 1881a; Jefferson, 1960; Ferrier 1890.

60. Ferrier,1874–1875.

61. Ferrier, 1875

62. Ferrier, 1875.

63. Ferrier, 1873.

64. Ferrier, 1886.

65. Ferrier, 1874–1875.

66. Graziano et al., 2002a, 2002b.

67. Ferrier, 1886.

68. Fritsch and Hitzig, 1870.

69. Hitzig, 1874. Quote translated by G. Krauthammer.

70. Ferrier, 1886.

71. Gross, 1998a; Glickstein, 1985.

72. Hitzig, 1874; Brazier, 1984.

73. Ferrier, 1876.

74. Horsley and Schäfer, 1883, 1888; Beevor and Horsley, 1887; Beevor, 1887; Brown and Sherrington, 1912, 1915.

75. Cheney and Fetz, 1985.

76. Leyton and Sherrington, 1916; Penfield and Rasmussen, 1950; Woolsey et al., 1952.

77. Walshe, 1943; Fulton, 1949b; Phillips, 1975.

78. Asanuma, 1975; Scott and Kalaska, 1997.

79. Georgopoulos et al., 1986, 1992.

80. Graziano et al., 2002a, 2002b.

81. Graziano et al., 2002a, 2002b; Graziano, 2006; Graziano and Aflalo, 2007.

82. Graziano, 2006; Graziano and Aflalo, 2007.

83. Strick, 2002.

II

Neuroscience and Art

I knew a five-year-old boy who had an insatiable thirst for art museums and art books. Among his favorites were Hieronymus Bosch and Peter Bruegel. In going over his Bosch and Bruegel books with him I was struck by images that looked like some weird version of a neurosurgical operation. In one case it looked like a flower was being extracted from the head and in another some stones were being removed. So, as described in chapter 5, I looked into this apparent "psychosurgery" in Renaissance art and its relation to trephination, a hoary practice whose history and current use are described in chapter 1.

Most of my own scientific work has been on the functions of inferior temporal cortex in object recognition. In the course of these studies I noticed that animals with damage to this part of the brain could tell left-right mirror images apart (e.g., p from q) as well as normal animals although they were severely impaired in telling apart other shapes. This led me to study left-right differences with Marc Bornstein, a colleague in the Princeton psychology department and an expert in both infant perception and art history. In chapter 6 we consider several general questions about right and left in science and art and to what extent differences between right and left are biological or social.

One of my history of neuroscience books has a Rembrandt painting of what looks like a brain dissection and another of an arm dissection. This piqued my curiosity to find out, as described in chapter 7, about Rembrandt's portraits of doctors dissecting corpses. Curiously, one of these paintings became a putative model for the final portrait of Che Guevara.

5

"Psychosurgery" in Renaissance Art

Hieronymus Bosch and other early Renaissance artists depicted "stone operations" in which stones were supposedly surgically removed from the head as a treatment for mental illness. These works have usually been interpreted either as portraying a contemporary practice of medical charlatans or as an allegory of human folly rather than a real event. Since trepanation for head injury and mental disease was actually carried out in Europe at this time, another possible interpretation of these works is that they are derived from a common medical practice of the day.

Bosch and *The Cure of Folly*

Hieronymus Bosch (ca. 1450–1515) is one of the most enigmatic painters of all time. Art historians have variously characterized this Flemish artist as a fanatic orthodox Christian, a satirical heretic, a pornographer, and a member of a secret black-magic sect (the "Adamites") worshipping the divinity of the sex act. His rich and flamboyant symbolism has been decoded (supposedly) in terms of alchemy, folklore and magic, various secret Christian sects, Freud and Jung, the Hebrew cabala, and the use of hallucinogenic drugs.[1]

Two things are clear about Bosch. One is that the plethora of conflicting interpretations reflects the imagination of art historians more than any solid evidence. The second is that Bosch had a twentieth-century mentality. His fantastic images have been viewed as anticipating Salvador Dali and the surrealist painters. Norman O. Brown said Bosch foreshadowed modern ideas of "therapeutic sexuality." Henry Miller claimed him as an inspiration to his own creativity. New interpretations of Bosch pour out and his images are found on such places as the covers of rock albums and novels on the holocaust.[2]

Perhaps the Bosch painting of most interest to neuroscientists is *The Cure for Madness (or Folly)*, also known as *The Stone Operation* (figure 5.1). This painting shows someone making a surgical incision in the scalp. The inscription has been translated as "Master, dig out the stones of folly, my name is 'castrated dachshund.'"[3] This is usually interpreted as reflecting a contemporary belief that folly, stupidity, and madness were due to stones in the head. "Castrated dachshund" was an epithet for a simpleton.[4]

The art-historical literature is replete with a large number of conflicting interpretations of the details of this painting such as the role of the two onlookers, the funnel on the surgeon's head, the book on the woman's head, the fact that a water tulip—not a stone—is being extracted from the head, the gibbet in the background, and other puzzling aspects. In spite of the disagreement on the meaning of the various apparent symbols in the painting, virtually all interpretations of the paintings fall into one of two classes. The first class views the painting as representing (and ridiculing) an actual practice whereby itinerant medical charlatans deceived people into believing that they could cure mental and "psychosomatic" symptoms by removing stones from the head.[5] Supposedly, the quack would make a scalp incision and then pretend to remove stones from the head. The second class of interpretation claims that there is no evidence at all for any such contemporary pseudo-medical practice.[6] Rather, the painting is viewed as an allegory of the extreme stupidity and gullibility of humans, a recurrent theme in Bosch.

Figure 5.1
Painting by H. Bosch, *The Cure of Folly* or *The Stone Operation* (ca. 1490, Prado, Madrid).
The text reads, "Master, dig out the stones of folly, my name is 'castrated dachshund'."
Yale University, Harvey Cushing/John Hay Whitney Medical Library.

After Bosch, there were a number of works, again usually Flemish, depicting the removal of stones from the head as a cure for madness and folly, including by Peter Bruegel, Jan Steen, Pieter Huys, Nicolaes Weydmans, and others (figures 5.2, 5.3, and 5.4). Following the two overall interpretations of the Bosch mentioned above, these later works have been interpreted either as depicting an actual common practice of medical quackery or simply as imitating Bosch's allegory of human stupidity (as each of these artists was clearly influenced by Bosch).[7] In both these art-historical interpretations of the depictions of "stone operations," the possibility that legitimate surgical operations on the head were actually performed to relieve symptoms was apparently quite inconceivable.[8]

TREPHINING

The oldest known surgical procedure is trephining, or trepanning, the removal of a piece of bone from the skull. It began in the late Paleolithic period and has been carried out in virtually every part of the world (see chapter 1). It is still used in the modern neurosurgical suite, in traditional Kenyan medicine and as an "alternative medicine" method of enhancing consciousness. Trephining has a strong and continuing tradition in Western medicine. It is described in detail in the Hippocratic work *On Wounds in the Head*, where it is indicated for various types of head injury. From the Renaissance until the beginning of the nineteenth century, trephining was widely advocated for the treatment of head wounds, particularly for depressed fractures and penetrating head wounds. It was also used, at least into the eighteenth century, for the treatment of epilepsy and mental disease.[9]

Roger of Parma (ca. 1170) wrote in his *Practica Chirurgiae (The Practice of Surgery)*:

> For mania or melancholy a cruciate incision is made in the top of the head and the cranium is penetrated, to permit the noxious material to exhale to the outside. The patient is held in chains and the wound is treated, as above under treatment of wounds.[10]

Figure 5.2

Engraving by Peter Bruegel the Elder, *The Witch of Malleghem* (1559. Malleghem was an imaginary village populated by the gullible, "mal" meaning crazy or foolish in Flemish. The witch is shown at the end of the table on the right holding up a stone she has just "extracted" from the fellow whose chin she still grasps. The stone had been provided by her lock-lipped assistant under the table. The advertising poster on the wall shows other stones she has removed and her surgical instrument. She is surrounded by others with stones in their heads waiting to be extracted. The figure in the foreground with a knife bound flat to his head represents sympathetic magic to draw out excess blood or bad humors, a custom that is said to have survived along the lower Rhine (Grabman, 1975; Klein, 1963). The verses under the print have been translated as "Folk of Mallegem, be of good cheer; / I Lady Witch, wish to be well-loved here . . . / I have come here to cure you / . . . Hurry on, everyone / If you've a wasp in your dome / Or are plagued by a stone" (Klein, 1963). Yale University, Harvey Cushing/John Hay Whitney Medical Library.

Figure 5.3
Painting by P. Huys (ca. 1562, Wellcome Institute Library; Schupbach, 1978).

loopt loopt met groot verblyden, Hier salmen twyf van hye snyden.

Figure 5.4
Engraving by N. Weydmans, *Operation for Stones in the Head*. The legend reads "Come, run, be filled with joy; Here we are cutting the woman of her stone" (Klein, 1963). Yale University, Harvey Cushing/John Hay Whitney Medical Library.

Robert Burton, in his still-in-print classic *Anatomy of Melancholy* (1652), similarly prescribed boring a hole in the head for melancholy:

Tis not amiss to bore the skull with an instrument, to let out the fuliginous vapors. . . . Guinerius cured a nobleman in Savoy by boring alone, leaving the hole open a month together by means of which, after two years' melancholy and madness, he was delivered.[11]

Thomas Willis, Professor of Natural Philosophy at Oxford, one of the founders of the Royal Society, and author of *Cerebri Anatome* (1664), the first comprehensive monograph on the brain (dealing with anatomy, physiology, and clinical neurology), noted that

Threatening, bonds or strokes were "Curatory" for madmen [but] Specifick Remedies such as St. Johns-wort as well as Chirurgical Remedies such as Trephining or opening the skull [were also indicated].[12]

Figure 5.5 is a 1573 woodcut showing a trephination in progress in a home operation. When the operation was moved into hospital settings in the beginning of the nineteenth century, the mortality rate was so high from the rampant infections characteristic of contemporary hospitals that trephination, for any reason, declined markedly until the introduction of modern antisepsis at the end of the century.[13]

BOSCH AND CONTEMPORARY MEDICAL PRACTICE

Cutting through the scalp into the cranium to remove a piece of bone, a practice known as trephination, was a standard surgical procedure during the periods in which the various depictions of "stone operations" were made such as those in figures 5.1 to 5.4. Furthermore, the procedure was

Figure 5.5
A 1573 woodcut showing a trepanation in progress. "Two people assist the surgeon, while a man warms a cloth, a women prays and two others watch" (Finger, 1994).

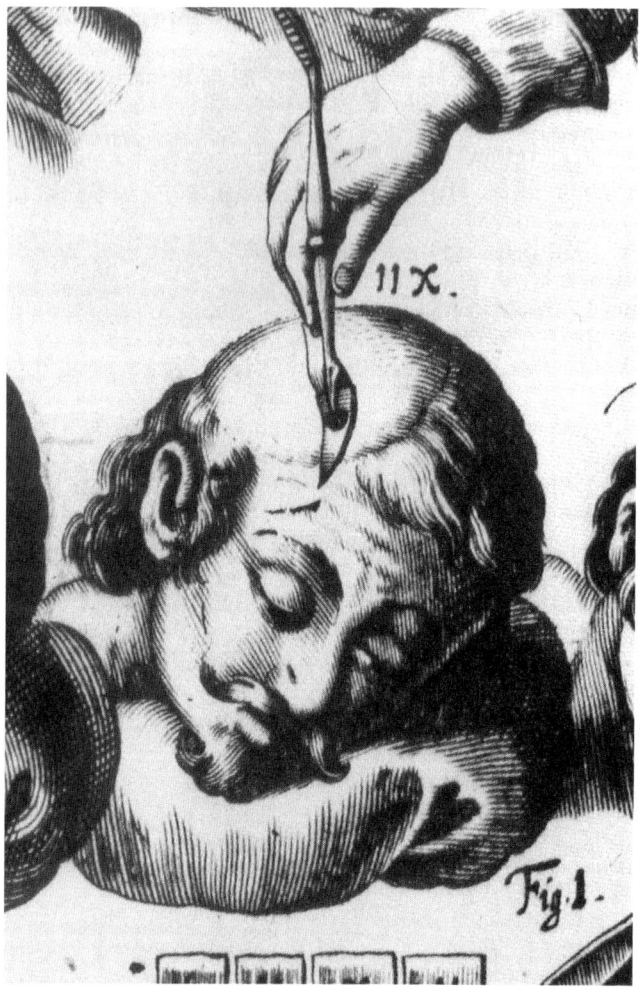

Figure 5.6
Detail of a figure from a 1653 work by J. Scultetus showing the start of a trepanation (Scultetus, 1655).

used to treat behavioral disorders as well as head injuries. Thus, it seems likely that Bosch, and the other artists who produced the various pictures of stone operations, knew of the existence of the actual contemporary medical procedure of trephination. Indeed, the details of their portrayal of the "stone operations" were sometimes close to the detailed instructional diagrams on trephination found in surgical handbooks such as Johannes Scultetus's *Armamentarium Chirurgicum* (1655; shown in figure 5.6).

Thus, whatever the abstruse symbolism in Bosch's *Cure of Folly*, whether he was ridiculing the church, the medical profession, trepanning, or all humanity; or whether it advocates some religious cult, some wild sexual practice, the advantages of trepanning, or nothing at all, it seems indisputable that the writing of art historians on this and similar works contains a great deal of folly. Apparently, unknown to many of these historians, Bosch's painting and derivatives by Bruegel and others were based on a very real medical practice of their time.

NOTES

This chapter is a version of an article previously published in *Trends in Neurosciences* (22: 429–431 [1999], "'Psychosurgery' in Renaissance Art").

1. Gardner, 1975; Cinotti, 1969; Gibson, 1973; Snyder, 1973; Harris, 1995; Delevoy, 1990; Bango Torviso and Marias, 1982; Bax, 1979.

2. See, e.g., Kosinski, 1999; *One Nation Underground* by the rock group Pearls Before Swine.

3. Cinotti, 1969.

4. Cinotti, 1969; Gibson, 1973; Harris, 1995; Bax, 1979; Schupbach, 1987.

5. Bango Torviso and Marias, 1982; Fry, 1946; Grabman, 1975; Klein, 1963; Mendena, 1969.

6. Gibson, 1973; Bax, 1979; Schupbach, 1987.

7. Grabman, 1975; Klein, 1963; Menden, 1969.

8. Gibson, 1973; Menden, 1969.

9. Gross, 1999b; Lisowski, 1967; Bakay, 1985; Saul and Saul, 1997.

10. Valenstein, 1997.

11. Burton, 1652.

12. Willis, 1683.

13. Dagi, 1997.

6

LEFT AND RIGHT IN SCIENCE AND ART

WITH MARC H. BORNSTEIN

How is the left hand different from the right hand? An asymmetric object, like a hand, can exist in two left-right mirror-image forms, or enantiomorphs, a phenomenon which has fascinated philosophers, cosmologists, and artists (figures 6.1 through 6.4). Psychologists and neurophysiologists have been particularly puzzled by the extreme difficulty children and other animals have in learning to distinguish left-right mirror images. In this chapter, we propose an explanation of why mirror images are so confusing. In the natural world almost all mirror images are actually two aspects of the same object, for example, the two sides of a face or a silhouette viewed from the front and back. Therefore a perceptual mechanism that treats mirror images as equivalent would be adaptive. The perceptual equivalence of mirror images only becomes maladaptive or confusing under very special conditions. One of these is learning an orthography containing mirror images such as b and d. Difficulty in learning to read may, in part, be due to difficulty in overcoming the normal tendency to treat mirror images as the same stimulus.

In the final portion of the chapter, we consider the effects of mirror-reversing a painting and, more generally, left and right in pictorial space. These effects suggest that some pictorial anisotropies, such as profile

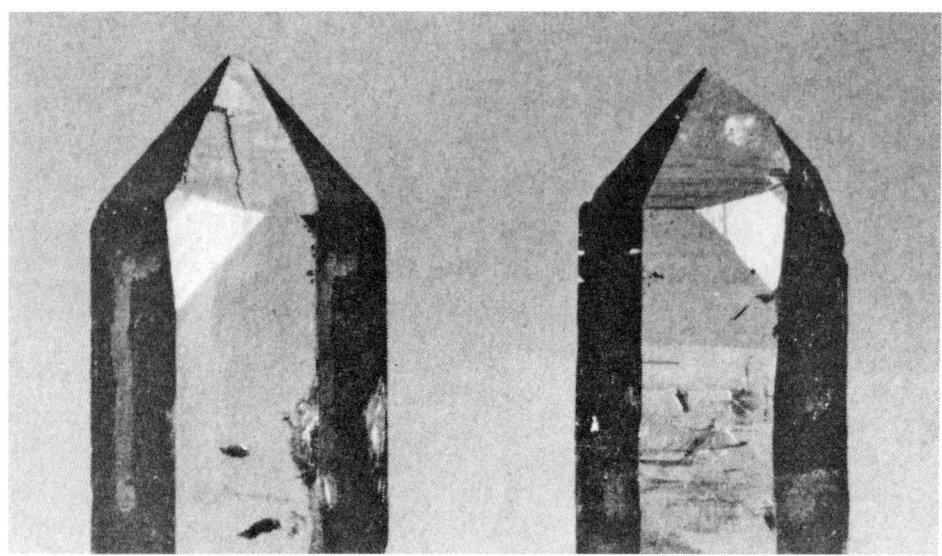

Figure 6.1
Quartz crystals. From the Nobel Prize address of V. Prelog, 1976.

Figure 6.2
Horns of Marco Polo's sheep (Thompson, 1969). Reprinted with the permission of
Cambridge University Press.

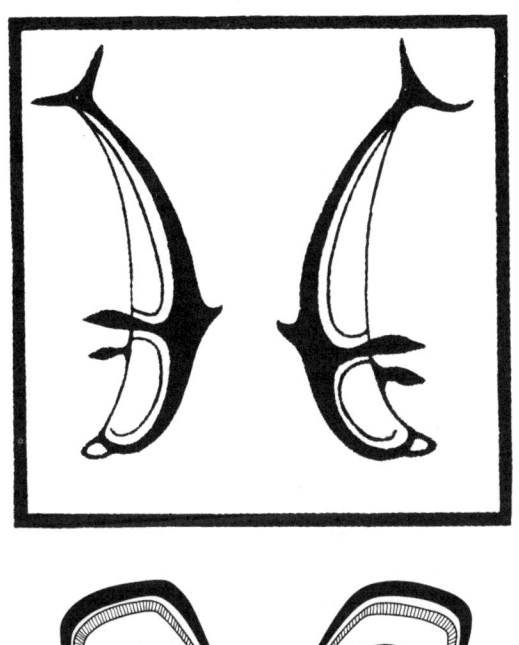

Figure 6.3
Floor pattern from the palace at Tiryns, Greece. Note that the two sides of a vertebrate are enantiomorphs (Swindler, 1929). Used by permission of Yale University Press.

Figure 6.4
Stylized bear from the Tsimshian Indians of the north Pacific coast. Note that the right and left sides of this or any other pattern with a vertical axis of symmetry are enantiomorphs (Boas, 1955).

orientation, reflect the influence of lateralized brain functions, whereas others, such as the tendency to look at a picture from left to right, are cultural conventions.

LEFT, RIGHT, AND COSMOLOGY

Is there a "left" and "right" in the universe? Newton said yes, and Leibnitz said no. Newton thought that the coordinates of space were absolute and "God-given." Leibnitz attacked this view and argued that left and right were "in no way different from each other."[1] Kant was puzzled by enantiomorphs for decades and this led him to side with Newton.[2] For Kant the difference between left and right mirror images was literally "inconceivable." They could only be distinguished through intuition, through the a priori structure of the mind, and therefore, for Kant, the a priori left-right structure of the universe.[3] Until very recently, however, cosmologists agreed with Leibnitz: the universe was symmetrical and right and left were arbitrary human conventions.

The problem was of strictly theoretical interest until many astronomers came to believe that there must be millions of inhabitable planets and therefore probably other beings at least as intelligent as humans. Distinguishing right from left then became of practical importance for developing a method of extraterrestrial communication. A digital code is clearly the best method, since it can be used to transmit pictures as well as messages. In order to decode pictures properly, it is necessary for the receiver to understand the instructions top and bottom, front and back, and left and right. Top and bottom can be described as "away" and "toward" with respect to the center of a planet. Front and back can be described as "near" and "far." But how does one describe left and right to an extragalactic listener? Describing left and right requires the sender to point to and the receiver to look at one side of an object, but this is impossible intergalactically.[4]

The solution to this problem came in 1957 from an experiment by Chien-Shung Wu at Columbia. Her experiment shook the very foundations of modern physics; parity had fallen. Madame Wu studied the emission of electrons from cobalt-60, a radioactive isotope of cobalt. Normally, electrons are emitted in all directions from cobalt-60, but, when it is cooled down near absolute zero ($-273°$ C) and placed in a strong magnetic field, the electrons would be expected to line up with the magnetic field and emerge equally from the two poles of the isotope nucleus. What Wu discovered was that more electrons came out from one side than the other of the otherwise uniform nucleus. Thus it was possible to label the poles of the magnetic field and therefore identify right and left in a consistent fashion. Right and left could now be given a meaning beyond human convention. Leibnitz was wrong: the universe is not symmetrical. Now, by describing Wu's experiment to our extragalactic audience we could tell them about left and right, and about which hand most humans write with.[5]

Why Are Mirror Images Confusing?

Like philosophers and physicists, children and psychologists have been confused about left and right (and cobalt-60 is unlikely to help them). The interest of experimental psychologists in left—right confusion began with the nineteenth-century physicist-philosopher Ernst Mach, who noted that "children constantly confound the letters b and d and also p and q...[and] ...adults too do not readily notice a change from left to right."[6] Since then, there has been an enormous volume of literature on letter reversal and more generally on the confusion of mirror images. A practical spur to this research has been the problem of reading disability because of its frequent association with letter reversal. The theoretical interest in mirror-image confusion is that it occurs in a variety of species such as octopus, fish, rats, and monkeys as well as human infants, children, and adults. The ubiquity of mirror—image confusion must therefore reflect something fundamental about how the nervous system processes visual information.[7]

Mach had suggested that the bilateral symmetry of the brain and body of the perceiver underlay mirror-image confusion and that, to the extent

---

that mirror images could be distinguished, "slight asymmetries...particularly in the brain" were responsible. In one version or another, essentially the same explanation of mirror-image confusion has been traditional in psychology ever since. The commonest version has been that of Samuel T. Orton, still the dominant name in the treatment of reading disorders.[8] Orton thought that because the two cerebral hemispheres were mirror symmetric, a visual stimulus on the retina would produce patterns of "neural excitations" in each hemisphere that would be mirror images of each other. Therefore, in order for a stimulus to be distinguished from its mirror image, the hemisphere with the veridical excitations would have to "suppress" or "dominate" the hemisphere with the mirror-reversed excitations. Mirror-image confusion, according to Orton, was ascribable to "incomplete dominance" of one hemisphere by the other.

Recently, Corballis and Beale have produced a new version of the Mach–Orton idea that mirror-image confusion derives from the symmetry of the brain.[9] Like Orton, they proposed that mirror-image confusion results from mirror-image representations of a stimulus in the two hemispheres. However, they realized that Orton's claim that an asymmetric stimulus would produce enantiomorphic patterns of stimulation in the visual areas of each hemisphere is false and that, in fact, the pattern in the two visual areas would be veridical and identical. Instead, they suggest that after the "neural excitation" is stored as a topographic memory in each hemisphere, the "memory representation" would mirror reverse when it transferred across the midline to the opposite hemisphere. (They assume that symmetrical points in each hemisphere are interconnected.) Thus each hemisphere would contain memories of the stimulus in both its veridical and its mirror form. According to their hypothesis, since every stimulus is stored in each hemisphere in its original and mirror-reversed forms, the organism would treat the two enantiomorphs as equivalent and therefore confuse them.

There are several serious difficulties with the various versions of the idea that mirror-image confusion derives from brain symmetry. One is pre-

---

sented by Gerstmann's disease. This disorder is characterized by extreme left-right confusion and follows damage to the left parietal lobe of the brain.[10] In this case an abnormally asymmetric brain increases left-right confusion rather than reducing it, as would follow from the Mach–Orton–Corballis hypothesis. Another argument against the role of brain symmetry comes from a comparison of children and animals. Both have severe difficulty in mirror-image discriminations, but the two hemispheres in children are functionally and anatomically asymmetrical, while in animals (at least below great apes) they appear to be relatively symmetrical.[11]

As mentioned above, Corballis and Beale realized that the initial representation of a stimulus in the two hemispheres would be veridical, not reversed as Orton had thought, and they suggested that the reversal would occur at a subsequent "memory" stage. However, this idea is also unsupportable by modern visual anatomy and physiology. The left visual half-field is represented in the right hemisphere and the right visual half-field in the left hemisphere as shown in figure 6.5. Moreover, there are multiple representations of the left half of space in the right hemisphere and of the right half of space in the left hemisphere.[12] However, the only anatomical connections between the representations of the left field in one hemisphere and the right in the other are between the representations of a narrow midline strip of the visual field.[13] This interconnection does not carry information about integrated visual patterns.[14] At this stage there are no longer any maps of visual space to be reversed. (However, the absence of evidence for an anatomical mechanism for interhemispheric reversal of memory traces is not a conclusive argument against the possibility of such a mechanism. After all, we learn and remember, and the anatomical bases of these phenomena are still totally obscure.)

More damaging for the Corballis and Beale hypothesis is that attempts with both humans and monkeys to demonstrate mirror reversal in interhemispheric memory transfer have failed.[15] That is, if a visual stimulus is presented to one hemisphere and the person or monkey is required to match it with a stimulus presented to the other hemisphere, the two hemispheres

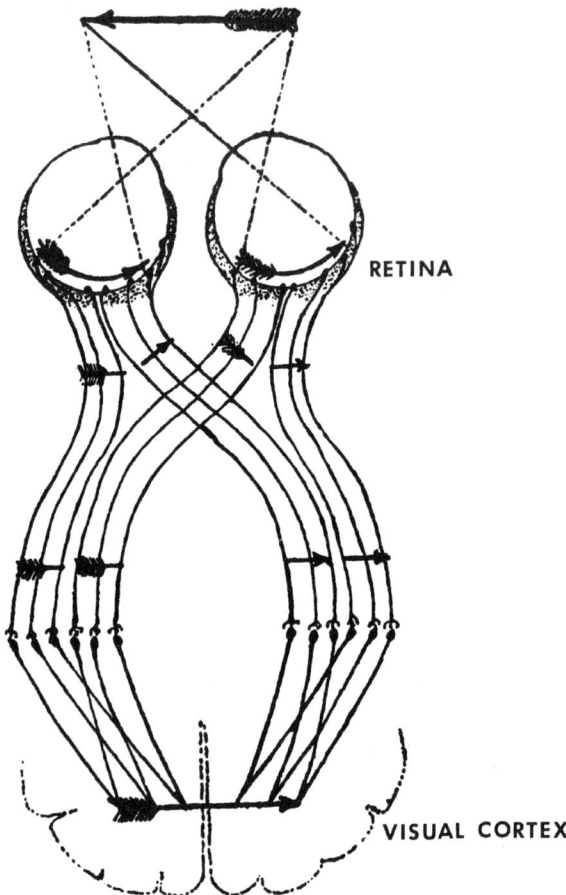

RETINA

VISUAL CORTEX

Figure 6.5

Diagram of the visual pathways from Ramón y Cajal's classic (1999). The labels have been added. Note that the optics of the eye reverse the image of the arrow on the retinae. The nerve fibers from each retina separate so that messages from the left half of each retina travel to the visual cortex of the left hemisphere, and the messages from the right halves travel to the visual cortex of the right hemisphere. Thus when the center of the arrow is fixated (as shown), information in the left half of space (the arrow head) goes to the right cortex, and information in the right half of space (the feathers) goes to the left cortex. Note further that the two cortical representations are not mirror-reversed with respect to each other.

invariably match the stimuli accurately. They do not equate the stimulus in one hemisphere with its mirror image in the other hemisphere, as would be predicted from the Corballis and Beale idea of "interhemispheric image reversal."

Finally, a general objection to both Orton and Corballis and Beale is that the existence of a topographic representation of a stimulus ("a picture") in the brain, whether veridical or reversed, does not constitute an explanation of perception. "Seeing is an interpretive process not a representational one."[16]

In summary, the Mach–Orton–Corballis hypothesis that mirror-image confusion derives from the symmetry of the body and brain does not work. Neither Orton's "incomplete dominance" nor Corballis's "memory reversal" fits the known facts.

AN EVOLUTIONARY HYPOTHESIS

If the symmetry of the body cannot explain mirror-image confusion, what can? We propose that the answer lies in the nature of the world and the evolution of the vertebrate visual system within it.

The selective pressure of evolution made it advantageous for the visual system to be able to perform certain types of visual processing while other types were unnecessary for survival. In the natural world there are never any mirror images that would be useful for an animal to distinguish. Indeed, with two exceptions, there are virtually never any mirror images at all. One exception is the two sides of a face or, more generally, the two sides of another bilaterally symmetrical animal. But here the two sides are aspects of the same thing, and it would be more adaptive to treat them as the same—not to distinguish them. Another exception is that a silhouette viewed from the back is the mirror image of the same silhouette viewed from the front. Again it would be adaptive to equate, not distinguish, these mirror images. In other words, we propose that the confusion of mirror images is not a "confusion" per se but an adaptive mode of processing visual

information. In the natural world the only mirror images that ever occur are aspects of the same thing and therefore need not be distinguished. Rather, it is adaptive to treat enantiomorphs as equivalent to each other. Instead of the "confusion" of mirror images we can speak of their perceptual equivalence.

To summarize our argument, mirror images are not confused because of the symmetry of the brain. Rather, mirror images are treated as equivalent because in the natural world mirror images are almost always aspects of the same thing.

Reading, Mirror Images, and Dyslexia

The perceptual equivalence of mirror images poses a problem only for humans and only when, for example, they have to learn to distinguish *b* from *d* and to write *s* and not its lateral mirror image in the Latin alphabet. Then children must learn to overcome the natural mode of processing that evolution built in to the brain, namely the perceptual equating of mirror images. On this view, mirror-image confusion is not unique to childhood except in the sense that it is as a child that a person usually has to overcome it. Thus, we would expect that nonliterate adults should show the same confusion as children. Corballis and Beale inadvertently confirm our prediction in reproducing the medieval woodcut shown in figure 6.6.

Reading is a complex skill, and children may have difficulty in learning to read for a great variety of reasons including poor instruction, antagonism between cultures of school and home, sensory and neurological disorders, impaired language development, emotional disturbances, and mental retardation. However, there is a small proportion of children for whom none of

Figure 6.6
Francis I Offers His Heart to Eleonore of Austria, by an anonymous French master, ca. 1536 (Corballis and Beale, 1976). Note that two *N*s and two *S*s are reversed in the print. Letters in the block for this woodcut should have been incised originally in reversed form. Shapiro (1970) observed that *S* and *N* are often reversed in early Medieval Latin inscriptions, as they are by children and unpracticed adults.

these conditions is present and yet they have severe and persistent difficulty in learning to read. Their condition is known as *developmental* or *specific dyslexia* or more simply "reading failure of unknown origin." Specific dyslexics are children of normal intelligence, vision, hearing, motivation, and oral language development who have inexplicable reading problems. These dyslexics can be taught to read eventually, but difficulty in reading, spelling, and learning foreign languages usually persists into adulthood.[17]

Reversal of letters in place (*b* and *d*), reversal of letters in a word (*on* for *no*), and failure to progress consistently from left to right are common errors in learning to read and write. However, they are even more common among dyslexic children. Although not universal, mirror reversals and left-right difficulties appear to be the principal distinguishing features of specific dyslexics other than their reading disability.[18] Can our explanation of mirror-image confusion also help explain specific dyslexia? Before suggesting how it may, it will be useful to review the difference between the left and right sides of the brain.

It is now well established that the two hemispheres of the human brain are each specialized for different psychological functions.[19] In virtually all right-handed people the left hemisphere is specialized for language. Therefore it is often termed the dominant hemisphere; but, strictly speaking, it is dominant only for language.[20] The right hemisphere is not just a nonlinguistic version of the left. Rather, the right hemisphere is specialized for and better than the left hemisphere at a range of perceptual functions that do not involve language. These include visual matching, memory for abstract designs, copying and drawing, face recognition, construction of block designs, and map reading. It seems likely, therefore, that the perceptual equating of mirror images is also a right-hemisphere function.[21]

Thus, specific developmental dyslexia may not reflect a deficit or dysfunction. Rather, we suggest it may indicate a relatively greater importance or dominance of the nonlanguage hemisphere (the right hemisphere in right-handers). This would result in slower acquisition of reading, not only because reading is a left-hemisphere (language) skill but also because mirror-image equivalence would be particularly strong or "good." A strong tendency to treat mirror images as equivalent would manifest itself in reversal of individual letters, in reversal of the order of letters in a word, and reversed scanning of phrases and lines: all classic symptoms of developmental dyslexia. According to this view, specific dyslexics should be superior to normal children on those perceptual, and perhaps artistic, skills that do not involve language, with one exception. The exception is in the discrimination of mirror images: On such tasks they would be much worse than normal children, since we are postulating that mirror-image equivalence ("confusion") signifies the predominance of the nonlanguage hemisphere.

The repeated reports that dyslexics are unusually artistic and better at drawing even before their reading disability manifests itself support this interpretation.[22] Unfortunately, there are very few studies that provide an adequate test of our hypothesis, because most are directed toward looking for deficits rather than superiorities among dyslexics. Furthermore, when perceptual abilities are examined, the results of perceptual tests dependent on language or on right-left discrimination are usually reported together with the results of tests that are not; we would expect dyslexics to be superior only on the latter ones. Finally, most studies of dyslexics include children who clearly have other difficulties in addition to learning to read, or they fail to compare these children with children of matched intelligence, age, and background. Thus it is not surprising that among studies of visuoperceptual abilities in dyslexia, most find dyslexics "unimpaired," a few find deficits, and only a very few find superiorities.[23]

If correct, our suggestion that specific dyslexics are superior at a variety of non-language-related perceptual skills may be relevant to techniques of teaching reading. In any case, to view dyslexics as potentially superior in so many other skills than reading and to encourage and develop this potential would certainly reduce the devastating emotional and social consequences of reading disability, which often are more crippling than the disability itself. It may be both more accurate and more helpful to view specific dyslexia as a difference rather than a deficit.

LEFT, RIGHT, AND ART

Gravity provides humans with an unambiguous basis for distinguishing up and down. Although Wu's cobalt-60 experiments revealed standards for labeling left and right, left and right are still confusing mirror images in physical space. Are differences in pictorial space confusing in the same way? Is there a left and right in pictorial art, and, if so, are they related to left and right in the brain?

Aestheticians have frequently asserted that left and right in a picture are absolutes.[24] Wölfflin and Gaffron have both emphasized, for example, that mirror-reversing a painting often drastically alters its meaning; thus they claimed that many of Rembrandt's etchings can be understood only by looking at their mirror image. For at least some paintings, this reversal effect is indeed striking. For example, mirror-reversing Janssens's *Reading Woman* (figure 6.7) clearly alters the composition.[25] Similarly, the effect of reversing Brueghel's *Parable of the Blind* is to change the impression of the painting so that "instead of tumbling into the ditch after their leader, [the blind men] seem to be pressing upon him."[26]

If, as aestheticians say, mirror reversal so changes the meaning of a painting, why have so many artists, from Raphael and Rembrandt to Munch, remained apparently indifferent to the reversal of their originals when reproduced as prints or tapestries?[27] And why, conversely, did a few, such as Dürer and Van Gogh, take great care to etch originals in their mirror image?

In fact, objective studies involving a number of observers and different paintings have lent little support to the generality of the claims of art historians that mirror-reversing paintings consistently changes the content or tone of the original. For example, when children and young adults were asked to tell whether they preferred the original or the mirror-image views of a series of classical (e.g., Raphael, Rubens, and Rembrandt) or modern paintings (e.g., Mondrian, Picasso, and Degas), preference for the original view was slight or nonexistent.[28] Similarly, in two other studies subjects could

Figure 6.7
Pieter Janssesn Elinga, *Die lesende Frau*. Bayerische Staatsgemäldesammlungen, Alte Pina-kothek Munich. Left: Veridical reproduction. Right: Mirror-image reproduction. The viewer's position relative to the perspective of the floor boards seems to change in these two versions. In the original, the viewer stands naturally at the woman's back looking over her shoulder; in the mirror-reversed version the viewer must change position to effect such an identification. In the veridical version the woman, in the left pictorial space, is important; in the mirror-reversal her slippers assume salience out of proportion, i.e., they are nearer and more conspicuous. Indeed pictorial depth is greater upon mirror-reversal. Gaffron (1950) suggested that these and other phenomenal differences arising from mirror-reversal evidence the existence of an unconscious central visual process or glance curve in pictorial perception. Thus, the slippers assume great prominence and appear closer because they stand at the head and in the path of the glance curve.

not accurately remember the left–right orientation of a set of representational pictures.[29]

A possible explanation for the failure of experimental psychologists to find the perceptual differences between paintings and their mirror reverses claimed by aestheticians might be that the psychological experiments involved collections of both symmetrically and asymmetrically organized compositions; by contrast, aestheticians exemplify their point with highly asymmetric paintings. Highly asymmetrical paintings, with marked perspective and lighting differences between the two sides (such as figure 6.7), clearly do alter on reversal. However, for naive adults viewing most paintings, as for children learning to read or for octopuses discriminating obliques, mirror images tend to be seen as equivalent. In all these cases the confusion or equivalence of mirror-reversed images represents a fundamental mode of visual analysis originating in organic evolution.

Independent of the difficulties people usually experience in telling left from right in nature or detecting mirror-image reversals of works of art, there appear to be verifiable differences associated with the left and right halves of physical space and pictorial space. Many aestheticians have suggested that the right visual field or objects located there tend to be perceived as heavier, larger, more distant, brighter, and more conspicuous, but less textured and less clear, while the left pictorial field or objects located there tend to have the opposite attributes.[30] Several of these claims have been verified experimentally. Specifically, under controlled conditions with a number of subjects, objects on the left have been found to appear closer and clearer and content on the right to appear heavier.[31]

BRAIN MECHANISMS OR ARTISTIC CONVENTIONS?

Asymmetries of pictorial space could arise from asymmetries of the brain or from cultural conventions. We suggest that both contribute to the anisotropy of art but in different ways.

As discussed above, the left hemisphere of the brain is principally concerned with the perception and production of linguistic material, and the right cerebral hemisphere is principally concerned with the perception and discrimination of nonverbal and spatial material. When visual input is strictly lateralized, and the eyes are not free to roam, the right half of visual space is processed primarily by the left hemisphere, and the left half of visual space is processed primarily by the right hemisphere (figure 6.5). Strict asymmetric input is, of course, not the case in artistic experience, since one does not just fixate the center of a painting. However, both Kinsbourne and Gur have demonstrated that visuospatial thinking and verbal thinking (or anticipation of these types of thinking) activate processes in the right and left hemispheres respectively and, at the same time, excite correspondingly lateralized eye movements.[32] Apparently, differential processing of information between the hemispheres is so basic as to maintain lateral biases even in the absence of strictly lateralized input, hence the importance of cerebral lateralization to art.

Anisotropies of visual space that are universal in humans probably reflect the influence of lateralized brain functions rather than cultural convention. One artistic asymmetry that appears to be universal in this way is profile orientation. Portraits are rarely full-face; one study found that of 1,474 painted portraits produced in Western Europe between 1500 and the present, a majority face leftward.[33] Similarly, right-handed children and adults of both sexes have a strong tendency (74% of 9,874) to draw profiles facing leftward.[34] This was found in the United States and Norway (where reading is from left to right), Egypt (where reading is right to left), and Japan (where reading is from top to bottom and right to left). By contrast, left-handed children were equally likely to orient their profiles in either direction, presumably reflecting the heterogeneous nature of left-handers and demonstrating that profile orientation is not a simple function of how the hand holds a pencil. Leonardo da Vinci, perhaps the most famous left-handed artist, preferred to draw right-facing profiles. An examination of

147

Figure 6.8
Leonardo da Vinci's *Heads of Girls and Men* ca. 1480 from the Royal Library at Windsor
(No. 12276 v).

the profiles in the Dover edition of Leonardo da Vinci's *Notebooks* shows that most of the profiles therein face right (figure 6.8).[35]

Profile orientation appears to be a function of laterality, not direction of reading, age, or sex. When a face is fixated centrally, the half of the face in the left visual field is processed by the right hemisphere (figure 6.5). As noted above, face recognition is largely a right-hemisphere function, and, when right-handed people look at the two halves of a front view of a face, the half of the face in the left visual field looks much more "like the person" than the other half.[36] Thus the tendency for portraits to locate profiles in the left visual field presumably reflects the fact that facial information there would be perceived more readily and accurately by the majority of people (i.e., right-handers). Similarly, as shown in figure 6.9, it is the expression on the half of the face in the left visual field that usually determines the right-handed viewer's impression of it.

In contrast to profile orientation, other aspects of visual anisotropy appear to reflect cultural conventions. Wölfflin, and others, have suggested that individuals typically enter a picture at the left foreground and proceed along a specified path or "glance curve" into the depth of the picture and over to its right-hand side.[37] They point out how this directional scan lends an aesthetic dimension of movement in graphic art. Movement from left to right in a painting is easier and faster, while movement from right to left is slower and perceived as having to overcome resistance. The former signals attack or approach; the latter withdrawal: in addition, the / diagonal is often associated with ascent and triumph, while \ is associated with descent and defeat. Poussin's *The Rape of the Sabine Women* (figure 6.10) epitomizes the use Western artists have made of these associations. The painting is a single dramatic frame out of an episode of the Roman legend. Figure 6.11 indicates the diagonals at work in the painting. The strong diagonal (figure 6.11, bottom) begins at the left with the entrance of the Roman fasces-bearer below (and Romulus above) and is continued in the Roman rout of the Sabines, which proceeds off the canvas to the right. The weak diagonal (figure 6.11, top), mainly focused in the father's futile attempt to save his

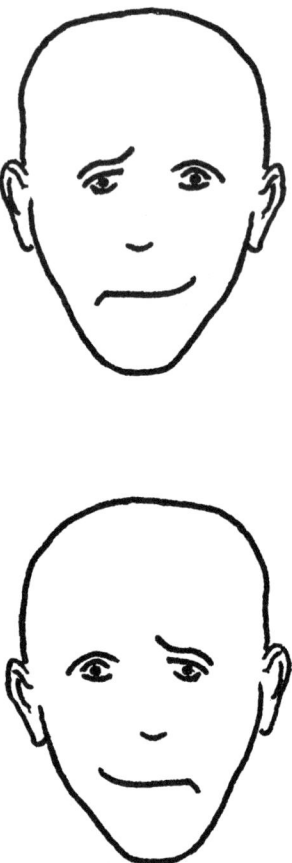

Figure 6.9
Stare at the nose of each face. Which looks happier? Jaynes (1976) found that most right-handers choose the bottom face, with the smile in their left visual field, presumably because the smiling side is processed by the right hemisphere on central fixation. Copyright © 1976, 1990 by Julian Jaynes. Reprinted by permission of Houghton Mifflin Harcourt Publishing Company.

daughter's honor in the lower right, is overwhelmed. These crossed diagonals represent opposing forces and express the theme of conflict and coun-
terconflict in a classical manner (the theme originally derives from the
metopes in the Parthenon); they inwardly organize a scene otherwise emotionally explosive and chaotic. Wölfflin believed that the left-to-right glance
curve represented a fundamental aesthetic vector. However, the glance
curve in Asian art appears to be in the opposite direction. Compare the directional selectivity of the Poussin with that evident in a detail from a Chinese handscroll—opened and read from right to left (figure 6.12). Here the
direction of viewing, first to the lower right and then toward the upper left,
is compelling.

Similar aesthetic vectors seem to operate in stage direction and audience expectation in the theater. According to Dean, the right stage (audience left) in Western theater is strong and elicits audience attention, so, as
the curtain rises at the beginning of an act, the audience can be seen to
look to the left front.[38] In Chinese theater, contrariwise, the important
positions are to the audience right. Thus the direction of the glance curve
in both painting and theater appears to be a cultural convention, presumably
related to the direction of reading, left to right in the West and right to left
in the East. As Gaffron says explicitly, "we 'read' a picture in a certain way
just as we read a page of a book."[39]

The term "glance curve" may be a misnomer, however, since studies
of eye movements across both Eastern and Western pictures do not reveal
glance curves in either direction.[40] Rather such studies suggest that the eye
roams over a picture in an arbitrary manner, only stopping to rest on salient
features. The glance curve may be some kind of covert cognitive scanning
with its direction set by reading habits. Or, alternatively, it may reflect a
cultural organizing principle implicit in graphic art. Whether in observers
or in pictures, conventions like the glance curve prevent representational
art from approaching excessive abstraction with its multiple and ambiguous
points of view.[41] If there were no such conventions about left and right to
show the way, artists and observers, like children, might be confused as to
how to view a picture.

Figure 6.10
The Rape of the Sabine Women by N. Poussin (1639). Metropolitan Museum of Art, New York.

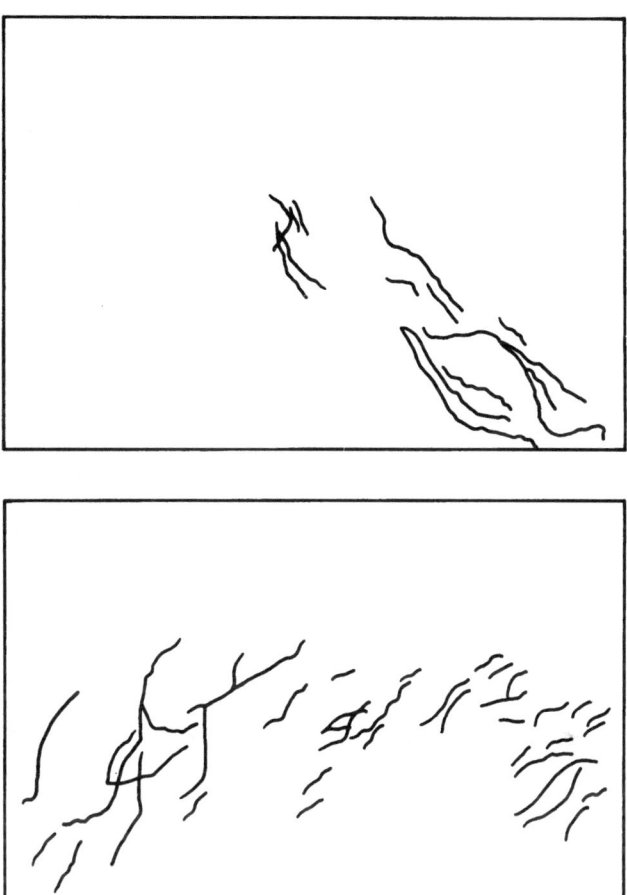

Figure 6.11
Weak and strong diagonal in Poussin's *The Rape of the Sabine Women*. Top: the "weak" diagonal (lower right to upper left). Bottom: The "strong" diagonal (lower left to upper right).

Figure 6.12
Twin Pines against a Flat Vista, by Chao Meng-Fu (1254–1322). Metropolitan Museum of Art, New York. Detail of a Chinese handscroll. Such handscrolls typically show dominant diagonals that follow the direction they are opened and read: right to left.

Postscript

In this 1978 paper we suggested that the well-known phenomenon of confusion of lateral mirror images (e.g., *b* and *d*) should not be conceived as a perceptual limitation but as an adaptive mode of processing: since in the natural world lateral mirror images are almost always views of the same object it would be adaptive to treat them as equivalent rather than distinguish them. That is, mirror-image "confusion" is not a confusion but an efficient perceptual constancy (unless you are a human learning to read).

This speculation has received considerable support since we first proposed it. First, visually responsive single neurons in inferior temporal cortex (crucial for visual pattern recognition) respond similarly to members of a lateral image pair,[42] an equivalence that may provide the basis of lateral mirror image confusion. Second, monkeys with lesions of inferior temporal cortex have great difficulty in distinguishing different patterns but they can distinguish members of a lateral mirror image pair as well as normals. This may be because lateral mirror images are easier to distinguish if mechanisms that specify their equivalence are lost. Since the animals with inferior temporal lesions have lost the mechanisms for equating lateral mirror images, they can learn to discriminate them. By contrast, normal animals treat mirror images as equivalent and therefore have difficulty in learning to tell them apart.[43] Third, in functional magnetic resonance images of human brains, lateral mirror images are treated as identical stimuli.[44] Similarly, in studies of visual search lateral mirror images are treated as equivalent.[45]

Another suggestion we made was that "dyslexics should be superior to normal children in those perceptual and perhaps artistic skills that do not involve language." In spite of the long list of great artists such as Leonardo and Picasso who were supposedly dyslexics,[46] there were actually very little adequate data on the artistic ability of dyslexics for the reasons we listed. Since then there has been at least one careful study of this question that did not have the weaknesses of previous studies. This involved the comparison of students at an elite university art school with comparative

---

students not studying art.[47] There was a much higher incidence of dyslexics among the art students, confirming our prediction.

Since this article was written there have been several good popular treatments of right and left in "brain, bodies, atoms and cultures"[48] and of left-handedness.[49]

NOTES

This chapter consists of an article published in the MIT journal *Leonardo* (11: 29–38 [1978], "Left and Right in Science and Art"), with minor edits and a new postscript. The co-author, Marc Bornstein, is now Senior Investigator and Head of the Section on Child and Family Research at the National Institute of Child Health and Human Development.

1. Leibnitz's critique of Newton occurred in his running controversy with Samuel Clarke, an English philosopher and divine acting as Newton's spokesman. See particularly Leibnitz's third letter to Clarke in Morris, 1951.

2. Handyside, 1929. For a discussion of the relation between mirror images and Kant's ideas of space see Remnant, 1963.

3. William James (1890) held a view somewhat similar to Kant, but he preferred the term "sensations" to "intuitions." "We can only point," he said, "and say here is right and there is left, just as we should say this is red and that is blue." However, James thought Kant incorrect in invoking the structure of space as a referent and thought the structure of the body to be sufficient.

4. This has been termed the "Ozma problem" by Martin Gardner, after Project Ozma, the first major attempt to develop a method of extragalactic communication. Ozma is the queen of the mythical kingdom in L. F. Baum's *The Wizard of Oz*. For discussion of this and many other problems of right and left in science, see Gardner, 1969; Adams, 1969; and Whyte, 1975.

5. For a nontechnical account of the fall of parity and its significance, see Wisner, 1965. There is a "meta-Ozma" problem: there may be pockets of antimatter in the universe and in these the results of the cobalt-60 experiment would be reversed. Thus, to communicate extragalactically about right and left it may also be necessary to know if our commu-

nicants are made of matter or antimatter. On the other hand, it may be possible to send a circularly polarized radio or light signal that would communicate the meaning of "right" and "left" even to an antimatter world without describing Wu's experiment.

6. Mach, 1914.

7. The classic animal and child studies are Sutherland, 1960, and Rudel and Teuber, 1963. Comprehensive reviews include Tee and Riesen, 1974, and Corballis and Beale, 1976. See also Bornstein, Gross, and Wolf, 1978.

8. Orton, 1937.

9. Corballis and Beale, 1976.

10. Critchley, 1953.

11. The following provide reviews and extensive bibliographies on hemispheric asymmetries: Schmitt and Worden, 1974—see particularly articles by Milner, Sperry, Berlucchi and Teuber; Kinsbourne, 1976, particularly articles by Levy and Trevarthen; Dimond and Beaumont, 1974; and Harnard et al., 1977.

12. See, e.g., Allman and Kaas, 1971; Gross, 1998a, figure 5.6.

13. Allman and Kaas, 1971; see, e.g., Brooks and Jung, 1973; Zeki and Sandeman, 1976.

14. Gross and Mishkin, 1977.

15. E.g., Hamilton and Tieman, 1973; Lehman and Spencer, 1973; Storandt, 1974; Corballis, Milner, and Morgan, 1971; Bradshaw, Nettleton, and Patterson, 1973.

16. Blakemore, 1973a.

17. Critical reviews of the dyslexia literature are in Money, 1962, particularly articles by Money, Saunders, and Benton; Critchley, 1970; and Benton, 1975.

18. Money, 1962; Shankweiler, 1963. The importance of mirror-image confusion to dyslexia is supported by the fact that specific dyslexia is apparently absent among Japanese children, and there are no mirror forms in Kana script, which is used for learning to read in Japan (Makita, 1968).

19. Schmitt and Worden, 1974; Kinsbourne, 1976; Dimond and Beaumont, 1974; Harnard et al., 1977.

---

20. The situation in left-handed people is not just the opposite of right-handed people. Some have language in the left hemisphere and visuospatial functions in the right, just like right-handers. Others are the reverse of right-handers in having language in the right hemisphere and visuospatial functions in the left. A third group of left-handers has both language and nonlanguage perceptual functions in each hemisphere. This heterogeneity of localization of function in left-handers probably reflects the fact that people are left-handed for several different reasons, such as heredity or some slight early trauma to the left hemisphere. See references in note 11 and Hecaen and Sauquet, 1971.

21. Indirect support for this possibility is the finding that right-handed people can read mirror-reversed words in their left visual field (projecting to their right hemisphere) more easily than in their right visual field (Harcum and Finkel, 1963).

22. E.g., Symmes and Rapoport, 1972.

23. Money, 1962; Shankweiler, 1963; Symmes and Rapoport, 1972.

24. Wölfflin, 1941; Gaffron, 1950; Oppe, 1944; Shapiro, 1970; Arnheim, 1974.

25. Wölfflin, 1941.

26. Gaffron, 1950.

27. Gaffron, 1950.

28. Swartz and Hewitt, 1970.

29. Ross, 1966; Gordon and Gardner, 1974.

30. Wölfflin, 1941; Gaffron, 1950; Arnheim, 1974.

31. Objects appear closer: e.g., Adair and Bartley, 1958; clearer: Dallenbach, 1923; Burke and Dallenbach, 1924; heavier: Levy, 1976.

32. Kinsbourne, 1974; Gur, 1975.

33. McManus and Humphrey, 1973.

34. Jensen, 1952a, 1952b.

35. Leonardo da Vinci, 1970. Why did Leonardo da Vinci mirror-write? Mirror-writing is common when a person either writes with two hands simultaneously or with the

nonpreferred hand, particularly without looking. This is presumably because if the muscles of the nonpreferred hand make the same movements as those normally made by the corresponding muscles of the preferred hand, the result will be mirror-writing. Perhaps Leonardo was taught with his right hand, and, because he was a left-hander, mirror-writing was the natural tendency when using his preferred left hand. Leonardo was hardly one to let convention alter his natural inclinations when writing notes to himself. The suggestion that his mirror-writing was a secret code hardly gives much credit to him or his contemporaries. In any case, as Luigi Boldetti showed, Leonardo "protected" many of his inventions by introducing an intentional error into his plans, such as an extra cog-wheel or an unnecessary ratchet (see Calder, 1970).

36. Gilbert and Bakan, 1973; Ellis and Shepherd, 1975; Young and Ellis, 1976.

37. Wölfflin, 1941; Arnheim, 1974; Swartz and Hewitt, 1970.

38. Dean, 1946.

39. Gaffron, 1950.

40. Buswell, 1935; Hess, 1965; Noton and Stark, 1971; Yarbus, 1967.

41. Landauer, 1969. Landauer asked undergraduates and professional artists for their preferences for one of four possible orientations of abstract paintings made by artists in the United States. Only 35% of the preferences were veridical.

42. Rollenhagen and Olsen, 2000; Baylis and Driver, 2001; Gross, 1978; Holmes and Gross, 1984.

43. Gross, 1978; Holmes and Gross, 1984.

44. van Zoest et al. 2006.

45. Biederman and Cooper, 1991; Fiser and Biederman, 2001.

46. West, 1997.

47. Wolff and Lundberg, 2002.

48. McManus, 2002.

49. Coren, 1992.

7

Rembrandt's *The Anatomy Lesson of Dr. Joan Deijman*

Rembrandt van Rijn's striking painting of a human brain being dissected by a headless figure, *The Anatomy Lesson of Dr. Deijman* (1657), may be the most famous portrayal of a neuroscience procedure (figure 7.1). It represents a curious combination of two genres of European painting: the group portrait and the historical painting, in this case an account of a public dissection. Such dissections served both educational and entertainment functions in seventeenth-century Holland.

The great popularity of group portraits was a unique Dutch phenomenon.[1] In sixteenth- and seventeenth-century Holland, there was no sovereign monarch or royal court and thus no royal patronage of the arts.[2] Furthermore, due to the rise of Calvinism, the Church no longer supported the arts. Thus, the principal large commissions available to artists were group portraits of the members of trading associations, hunting clubs, and guilds or other civil institutions. As the subjects were usually the organization officers or "board of regents," these group portraits are often termed "regents' portraits." *The Night Watch* (1642), one of the most famous of Rembrandt's masterpieces, was such a commissioned group portrait of a company of Civic Guards in which the amount of money each subject contributed determined his prominence in the painting.[3]

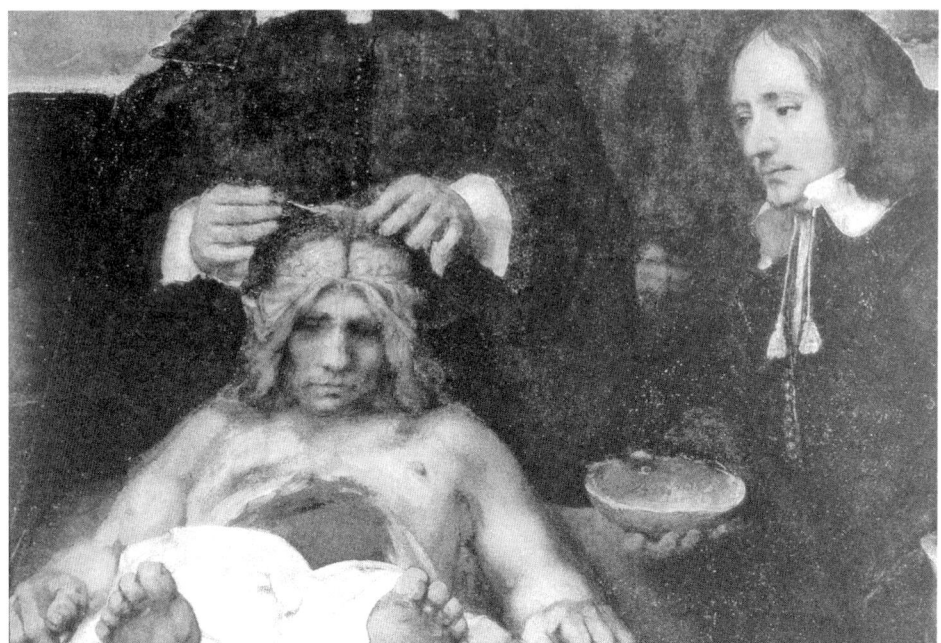

Figure 7.1
Rembrandt, *The Anatomy Lesson of Dr. Joan Deijman* (surviving fragment of original), 1656. Amsterdams Historisch Museum.

One type of regents' portrait was the "anatomy lesson," a group portrait of the leading members of a surgical guild (or, more specifically, those members who contributed to the cost of the painting).[4] Before Rembrandt, the figures in these paintings were usually gathered rather stiffly and artificially around a skeleton, skull, or body being autopsied (figure 7.2). In Rembrandt's hands, however, the "anatomy lesson" developed beyond a group portrait to become a more or less accurate account of a significant historical event in the life of the bourgeois Dutch community, namely the public dissection of an executed criminal.

Public Dissections as Theater

The spectacle of public dissections in front of large audiences of both medical professionals and laymen began in the early Renaissance medical schools of Italy and had become common throughout the continent by the middle of the sixteenth century.[5] In Holland, they were an elaborately regulated public ritual in the major cities. Since each city usually authorized only one such public dissection each year, it became a major event in the Dutch social, educational, and entertainment calendar; it went on for three to five days following the execution.[6]

The dissections were conducted by a leading surgeon in the community who had been appointed the city *praelector chirurgic et anatomie*. They were held in the winter to retard putrefaction of the bodies and were conducted in special anatomical theaters that held 200 to 500 spectators (see figure 7.3).[7] The affairs were evening events, illuminated by scented candles and often accompanied by flute music. The rival professional groups, the physicians and surgeons, sat separately from each other and from the lay public. Everyone was charged admission. The proceeds were used not only to pay the fee of the praelector but also for food, drink, and tobacco at the major banquet of the Guild of Surgeons. The banquet was followed by a torchlight parade.[8]

Figure 7.2
N. Eliasz, *The Anatomy Lesson of Dr. Johan Fonteijn*, 1626. Amsterdams Historisch Museum.

VERA ANATOMIÆ LUGDUNO-BATAVÆ CUM SCELETIS ET RELIQVIS QVÆ IBI EXTANT DELINEATIO.

ARCHIVVM INSTRU:
MENT. ANATOMICOR.

Figure 7.3

W. Swanenburgh (1581–1612) after J. C. Woudanus (ca. 1570–1615), *The Anatomical Theater in Leiden*, 1610. Amsterdam, Rijksmuseum. Rembrandt's drawing of a human skeleton riding a skeleton horse, *Skeleton Rider* (ca. 1655, Darmstadt Museum) is believed to be of the skeleton in the upper right, or possibly of a similar exhibit reported in the anatomical theater in Amsterdam. The drawing, in turn, is thought to be closely related to Rembrandt's The *"Polish" Rider*, ca. 1655. New York, Frick (Held, 1991).

In ordinances of 1605 and 1625 regulating the dissections in Amsterdam, the audience was explicitly forbidden from talking or laughing during the dissections. They could ask questions as long as they were of a "decent and serious nature." Body parts such as the heart, kidney, and liver (the *membra naturalia*) were passed among the audience but stiff fines were in place to ensure their return.[9] At least until the seventeenth century only male bodies were used.

These rare dissections were particularly valuable for the physicians and surgeons in the audience since anatomy was then viewed as the fundamental basis of medicine and surgery. In addition, contemporary accounts stress the educational value of the dissections for the general audience, for example, in demonstrating "the secrets of nature revealed by God." They also continued the hoary practice of discouraging crime by mutilation of the criminal's body after death. Finally, they were also very good theater.[10]

Some historians have emphasized the more general scientific, artistic, and cultural roles of the anatomical theaters.[11] At that time, Holland, unlike Britain, France, and Italy, had neither scientific societies nor scientific journals. Thus the anatomical theaters served as important venues for scientists to meet and discuss their work. Lectures on medical and other topics were routinely scheduled there. The anatomical theaters usually included attached libraries, museums ("natural history cabinets"), and even botanical gardens. Major paintings were exhibited, particularly, of course, "anatomy lessons."

THE ANATOMY LESSON OF DR. NICOLAES TULP

Rembrandt painted two anatomy lesson group portraits, both of members of the Surgeons' Guild of Amsterdam. The first, painted in 1632, was *The Anatomy Lesson of Dr. Nicolaes Tulp* (figure 7.4). It was commissioned by Tulp, the city praelector, and paid for by those portrayed in it (except for Aris Kint, the cadaver, who had been hanged for robbery with violence).[12] Although in the form of the usual group portrait, it was actually a strikingly

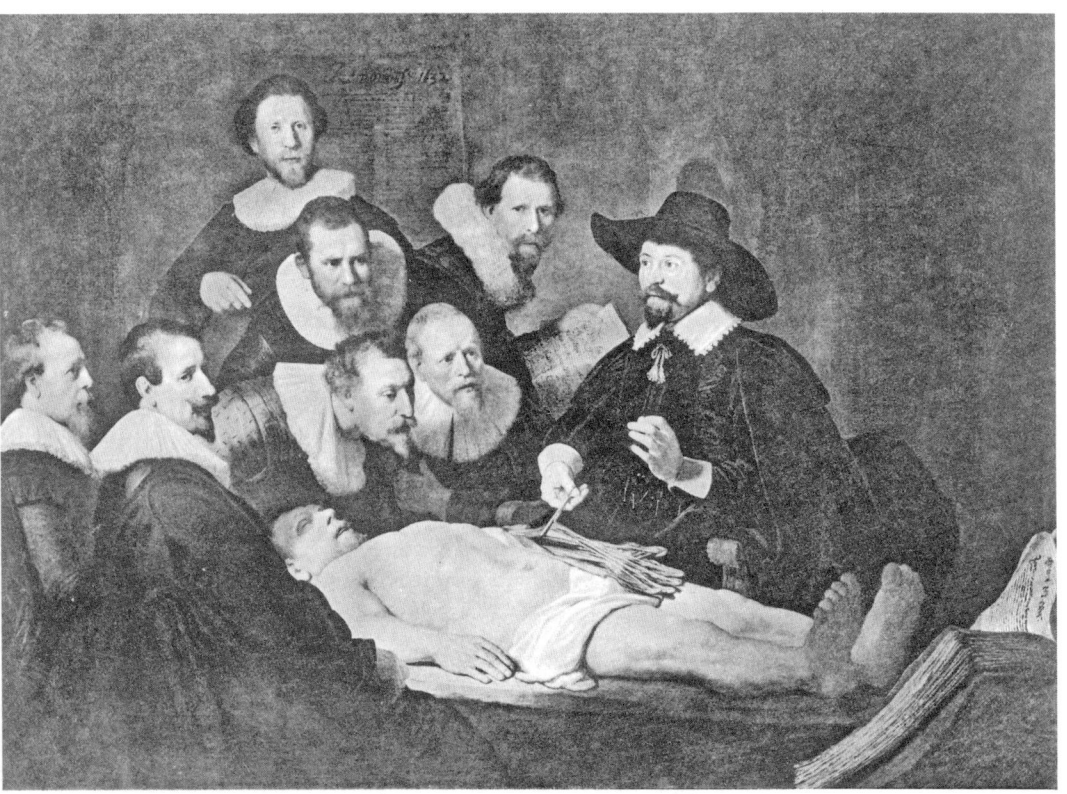

Figure 7.4
Rembrandt, *The Anatomy Lesson of Dr. Nicolaes Tulp*, 1632. The Hague, Mauritshuis Museum.

original artistic triumph. For the first time, Rembrandt dramatically emphasized the dissection rather than just the dramatis personae. The lecture-dissection is shown as the dramatic event it must have been. The portraits do not look "pasted on" as was usually the case before (figure 7.2); rather they are portrayed as individuals, with considerable variation in how and how much each is paying attention to the dissection.

In spite of its artistic superiority, the painting was typical of previous "anatomy lessons" in deviating considerably from an accurate account of the real event.[13] At the actual dissection, the guild members, other than the lecturer and perhaps his assistant, would have been in the front rows of the audience rather than on the stage around the lecturer. Tulp is shown starting by dissecting an arm when, in fact, the body cavity was always opened first and the limbs were usually not reached until the second day. Finally, Rembrandt's rendering of the anatomy of the arm is rather distorted and inaccurate.

The painting was Rembrandt's first group portrait and it was an immediate success. It established him as a major portrait painter, leading to many portrait commissions. In fact, medical professionals were "among the most faithful of Rembrandt's patrons throughout his life."[14]

The fame of *The Anatomy Lesson of Dr. Nicolaes Tulp* went beyond the world of art history. As was first pointed out by the English critic John Berger, it was the model for the picture of the murdered Che Guevara that sprung from the first pages of newspapers around the world: the Christlike figure lying half-naked and surrounded by Bolivian officers and soldiers, the commanding officer mimicking the stance of Dr. Tulp.[15]

THE ANATOMY LESSON OF DR. JOAN DEIJMAN

Twenty-four years later, Rembrandt painted his second and final anatomy lesson. Now the praelector was Dr. Joan Deijman, Tulp's immediate successor. This painting, originally measuring 245 by 300 centimeters, was badly damaged in a fire in the eighteenth century and only the central portion

(113 × 135 cm) of the lower half survived. This fragment then disappeared and was rediscovered in England in the nineteenth century, and badly slashed in the twentieth.[16]

After the painting was completed, Rembrandt made a sketch of it for the design of its frame. On the basis of this sketch, of contemporary accounts of the intact painting, and of Rembrandt's other portraits, the original painting has been reconstructed (figure 7.5).[17] Whereas the original consisted of the portraits of eight spectators in addition to Dr. Deijman, his assistant, and the cadaver, in the fragment only the cadaver, the assistant, and a headless Dr. Deijman survive (figure 7.1). The body was that of Joris Fonteijn, who had just been hanged for his "habitual criminality."

The portrait is a more accurate account of the standard public dissection than Rembrandt's earlier painting. The dissection of the viscera has been completed first and Deijman has removed the top of the skull (which his assistant is holding), has flapped back the dura, and is presumably about to start the usual next step, horizontal sections of the cerebrum. It is not clear whether Rembrandt actually sketched this or any other brain dissection from life. However, the view of the brain and the flapped-back tissue is virtually identical to plate 67:2 of Andreas Vesalius in his *On the Fabric of the Human Body* (1543),[18] even in regard to the expression of the mouth. It is thus almost certain that Rembrandt depended heavily on this Vesalius figure, whether or not he also made his own observations.

The overall design of the painting also seems to have been influenced by the famous title page of Vesalius's great work.[19] In the center of that woodcut, Vesalius is shown dissecting a human corpse in front of a Palladium-like stage.[20] Although most of Rembrandt's other multiperson paintings are asymmetrical, the principal features of this painting, the body and the dissector, are placed in its very center. (Even Christ does not usually get this treatment in Rembrandt.) The stagelike structure with which Rembrandt surrounds his picture may also be derived from the stage in the background of the Vesalius frontispiece; there does not seem to have been a similar stage in the Amsterdam anatomical theater.

Figure 7.5
Rembrandt, *The Anatomy Lesson of Dr. Joan Deijman*, computer montage reconstruction
by N. Middelkoop and T. Wolzak (Middelkoop, 1994) Amsterdams Historisch Museum.
The missing portraits are taken from other Rembrandt paintings.

Artistically, *Dr. Deijman* is considered an even greater masterpiece than *Dr. Tulp*.[21] The radically foreshortened body is particularly dramatic, leading the viewer's eye from the confronting feet, across the open viscera to the brain and the scalpel. (This arrangement of the body is considered to be derived from Montegna's *Dead Christ* in Milan.) The eyes are just enough in shadow to threaten to stare directly at the viewer. The blackish toes and lips, the yellow skin tone, and the rigor mortis are all of a reality never before seen in an "anatomy lesson." When the great English painter and critic Sir Joshua Reynolds saw the full original in 1781 he commented, "There is something sublime in the character of the head which reminds me of Michael Angelo; the whole is finely painted, the colouring much like Titian."[22]

Today, the original functions of "anatomy lesson" paintings are fulfilled by group photographs usually posed in front of the organization's building or meeting place. The lay functions of the public dissection, namely entertainment, voyeurism, and education, are largely carried out by television.

***

POSTSCRIPT

This article as originally published did not have the space to deal with three other aspects of the "anatomy lesson" paintings of Rembrandt. The first is who were Drs. Tulp and Deijman, the second is how accurate are the dissections shown in the anatomy lessons, and the third is who were the subjects of the dissections.

DR. TULP AND DR. DEIJMAN

If Dr. Nicolaes Tulp (1593–1674), born as Claes Piterszoon, had not been immortalized by Rembrandt, he would still be a paragraph or two in a (large) history of European medicine.[23] His major work was the four-volume *Observationum Medicarum* (1641 and new editions continued until

1739). It was primarily a set of over 200 cases covering what today are a variety of specialties. It was called the "Book of Monsters" because many of the cases were quite bizarre. It included illustrations and descriptions of some of the strange animals that the Dutch East India Company was bringing home. One of these seems to have been the first ape ever brought back to Europe and described (figure 7.6).[24]

Among the significant observations that medical historians have mined from this work were the first description of spina bifida (a developmental spine disorder) and of "cluster headache" (a brief recurrent unilateral pain involving the temples, neck, and eye), and cases of contralateral hemiplegia and posttraumatic amnesia and of a variety of gastrointestinal and urological disorders. His anatomical discoveries include a valve of the intestines known as "the valve of Tulpus" and the first description of human lacteal vessels. Tulp was a man of this time: Hippocrates and Galen were quoted much more often than any contemporary sources.

In seventeenth-century Holland physicians were supposed to teach and guide the lower-class surgeons and not do any actual surgery themselves. However, Tulp described several successful trephinations for head injury, which he appears to have carried out himself.

Tulp became successful and wealthy and entered politics, eventually holding the post of burgomaster (like a mayor) four times. Curiously, in the 1980s a fitness report on the first Dutch settlers on Manhattan Island, signed by Tulp, was found in the archives of the Dutch colony in the New York public library.[25]

Dr. Johannes Deijman (1620–1666) succeeded Tulp as the Amsterdam *praelector chirurgic et anatomie* in charge of public dissections. He does not seem to have published anything.

THE ACCURACY OF THE DISSECTIONS

There has been a long debate on the accuracy of the forearm dissection in *The Anatomy Lesson of Dr. Nicolaes Tulp*. One issue was whether it was the

Figure 7.6
The title page of Tulp's *Observationum Medicarum* (1641) illustrating four of his "cases."
The top drawing shows a blacksmith who removed his own large kidney stone with the
help of his young son. The left figure shows a man with an inserted wooden tube to drain
accumulated ascites fluid. The right figure is a woman with multiple ovarian tumors. The
bottom figure is the "orang-utang," actually a chimpanzee, later reproduced by Tyson
(1699) and by Huxley (1863). From Dudok van Heel, 1998.

arm of another body, since usually the viscera are exposed first and some thought the joining of the arm to the body was odd. Another issue was the accuracy of the dissection itself, partly because some relevant landmarks are not clearly seen. However, recent investigations—including one that compared Rembrandt's depiction to a "dissected left forearm of a Dutch male cadaver" and not just to drawings in anatomy textbooks—seem to indicate that Rembrandt was astonishingly accurate in his depiction of the anatomy of the arm.[26]

In 2000 the neuroanatomist Laurence Garey and the historian of medicine William Schupbach had a more critical attitude to *The Anatomy Lesson of Dr. Deijman*.[27] First, they claimed that Rembrandt had eliminated "at least a cubic foot" between the body's head and foot. In a more serious criticism, they claim that Dr. Deijman had cut the falx (the dura between the two hemispheres) and turned it around to show its crescent-shaped side to the viewer. *Falx* is Latin for scythe, often carried by the personification of death. They suggest this was some kind of a pun by Deijman or Rembrandt.

THE SUBJECTS OF THE DISSECTIONS

Tulp is dissecting the body of Aris Kint, a "habitual" petty criminal who found prison life so awful that he apparently tried to stab a guard in order to get a death sentence—which he succeeded in doing.

Deijman is dissecting the body of Joris Fontejin, who, with his girl-friend, was trying to rob a textile shop when he injured one of the shop assistants who tried to stop him. That Fonteijn was carrying a pistol may have contributed to the severity of the sentence.[28]

REMBRANDT AND CHE GUEVARA

In this chapter I noted the suggestion of the art critic John Berger that Rembrandt's *The Anatomy Lesson of Dr. Nicolaes Tulp* (figure 7.4) was the model for the famous Christlike picture of the murdered Che Guevara

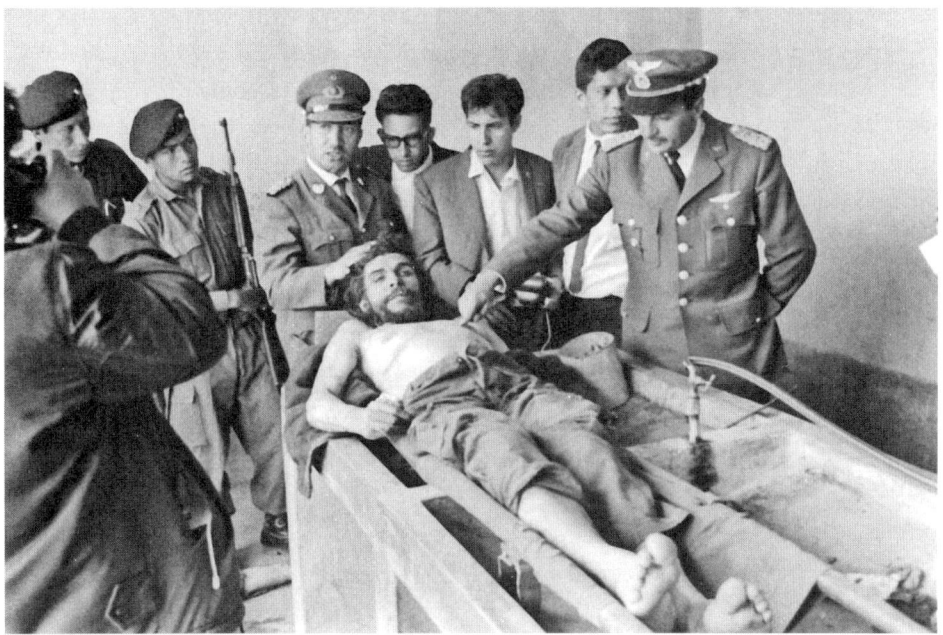

Figure 7.7
The Corpse of Che Guevara, Freddy Alborta, 1967. After capturing and executing Che in 1967, before burying him in a secret site, the executioners posed for this group photo with the body to demonstrate that Che was really dead. Later the body was exhumed and now rests in an elaborate churchlike memorial in Santa Clara, Cuba. This photograph has been named as one of the "10 Photographs that Changed the World" (www. arkitectrue.com/10-photographs-that-changed-the-world, accessed October 1, 2008). Used by kind permission of the Estate of Freddy Alborta Trigo, La Paz, Bolivia.

(figure 7.7). Indeed the organization of the two works is certainly strikingly similar. However, the Bolivian photographer who took the picture, Freddy Alborta, has insisted in a film about the photograph, *El Día Que Me Quieras* (Leandro Katz, 1999), that the similarity of the two was coincidental and that he was unfamiliar with the Rembrandt. Apparently, the photographer remained largely unknown to the public until Katz's film about him.

NOTES

This chapter (excepting the postscript) was originally published in *Trends in Neurosciences* (21: 237–240 [1998], "Rembrandt's *The Anatomy Lesson of Dr. Joan Deijman*").

1. Middelkoop, 1994; Rosenberg, 1968.

2. Schama, 1987.

3. Rosenberg, 1968.

4. Middelkoop, 1994; Rosenberg, 1968; White, 1984; Heckscher, 1958; Hansen, 1996.

5. Singer, 1957; Gross, 1998a.

6. Middelkoop, 1994; White, 1984; Heckscher, 1958; Hansen, 1996; Rupp, 1992; Cazort et al., 1996.

7. Held, 1991.

8. Middelkoop, 1994; Heckscher, 1958; Hansen, 1996; Rupp, 1992; Cazort et al., 1996.

9. Heckscher, 1958.

10. Hansen, 1996; Rupp, 1992; Bal, 1991.

11. Hansen, 1996; Rupp, 1992.

12. Heckscher, 1958.

13. Heckscher, 1958; Hansen, 1996.

14. White, 1984; Clark, 1978.

15. Weschler, 2006.

16. Middelkoop, 1994.

17. Middelkoop, 1994.

18. Saunders and O'Malley, 1950

19. Gross, 1998a, figure 1.8.

20. Saunders and O'Malley, 1950.

21. Rosenberg, 1968; White, 1984.

22. White, 1984.

23. The most detailed account of the life, work, political career, and social context of Tulp is the lavishly illustrated Dudok van Heel, 1998. Other recent papers on Tulp are Kruger, 2005; Simpson, 2007, and Mellick, 2007.

24. Tulp called it an "orang-outang," and his drawing of it (figure 7.6) was reproduced in Tyson's (1699) founding primatology work and in Huxley's classic *Man's Place in Nature* (1863) where he noted, correctly, "It is plainly a Chimpanzee."

25. Dudok van Heel, 1998; Kruger, 2005; Simpson, 2007; Mellick, 2007.

26. IJpma et al., 2006; Mills, 1989; Alting and Waterbolk, 1982; Lindeboom, 1977; Jackowe et al., 2007.

27. Reed, 2000; Anonymous, *Imperial College Reporter*, 2000 http://www.imperial.ac.uk/P2496.htm (accessed February 25, 2009).

28. Dudok van Heel, 1998.

III

Scientists Who Were "Before Their Time"

Often in the history of science, when a scientist's ideas or interpretations are too novel, they are rejected or simply ignored. Of course, in most of these cases the new ideas turn out to be wrong. Much more rarely, these ideas become accepted as major insights decades or even centuries later: they had been "before their time." This part of the book examines five rather different cases of discoveries or ideas that later became important in neuroscience but were not appreciated or understood in their time. In each case we consider why the scientists' contemporaries ignored their work.

The first case is Claude Bernard, the most famous French scientist of the nineteenth century and arguably of all time. Although his experimental discoveries were immediately accepted, his central theoretical idea that the constancy of the internal environment is necessary for the development of a complex nervous system had no meaning until about 50 years later.

Bartolomeo Panizza is a rather different case. In the middle of the nineteenth century he produced the first clear experimental evidence for localization of function in the cerebral cortex; specifically, he located a visual area in the posterior cerebral cortex in a variety of animals including humans. Yet this work went unrecognized until after the establishment of cortical localization by Broca (1861) and Fritsch and Hitzig (1870) and then the rediscovery of a posterior visual area by Munk (1881).

The third case is that of Joseph Altman, who in the 1960s overturned the dogma, as old as modern neuroscience, that no new neurons are made in the brains of adult mammals. Although prominently and repeatedly published and even replicated, his findings were ignored until about thirty years later.

The fourth case, Donald Griffin, is someone who was ahead of his time in two ways. First, he discovered (with Robert Galambos) something previously inconceivable, namely that bats could navigate with an amazingly accurate sonar-like system, echolocation. Second, in the face of much skepticism and even ridicule he restored the study of animal consciousness to experimental science, from which it had been expelled by the rise of behaviorism. Griffin was my first scientific mentor, and this memoir of his life and work is written in appreciation of that role.

The final case involves several investigators. In the 1960s Jerzy Konorski hypothesized the existence of neurons whose activity would code for complex visual percepts such as a face. Jerry Lettvin speculated about a cell that might code the percept of your grandmother, hence the term *grandmother cell*. Subsequently, my colleagues and I found cells that would fire selectively to facial images, but it took another twelve years before anyone tried to repeat our results (and they were successful) and another seven years until similar mechanisms were sought in humans.

In my previous *Tales in the History of Neuroscience* I recounted a more extreme case of neglect. In the eighteenth century, when the cerebral cortex was usually considered to have only a "rind" function, Emanuel Swedenborg (1688–1772) argued for sensory and motor function and even a type of "quasi-neuron" theory. Yet his theories remained unknown until the twentieth century, by which time many of his ideas had been confirmed.

Resistance by scientists to new ideas and discoveries may not be rare. Herman von Helmholtz (1821–1894), writing to Michael Faraday (1791–1867), noted:

The greatest benefactors of mankind usually do not obtain a full reward during their lifetime, and . . . new ideas need more time for gaining general assent the more really original they are. (Barber, 1961)

Max Planck (1858–1947) commented about some new ideas in his doctoral dissertation:

None of my professors at the University had any understanding for its contents. . . . I found no interest let alone approval, even among the professors who were closely connected with the topic. Helmholtz probably did not even read the paper at all. . . . *A new scientific truth does not triumph by convincing its opponents and making them see the light, but rather because its opponents eventually die and a new generation grows up that is familiar with it.* (Barber, 1961; italics mine)

8

CLAUDE BERNARD AND THE CONSTANCY OF THE INTERNAL ENVIRONMENT

Claude Bernard (1813–1878) was the founder of modern experimental physiology and one of the most famous French scientists of all time (figure 8.1). Today, his fame rests primarily (if not entirely) on his idea that the maintenance of the stability of the internal environment (*milieu interieur*) is a prerequisite for the development of a complex nervous system.[1] In Bernard's time, his many experimental discoveries in physiology were widely recognized and he received virtually every honor possible for a scientist in France. Yet, his conception of the internal environment had no impact for more than 50 years after its formulation. In this essay, after his life and major work are summarized, some reasons both for the delay in the recognition of this idea and for its subsequent importance to the physiology of the first quarter of the twentieth century are examined.

LIFE AND MAJOR WORK

Claude Bernard came from poor peasant stock in the Rhone Valley. At the age of 19, after some nonscientific education, he was apprenticed to a local pharmacist. Bernard was more interested in writing plays, however, and set out for Paris in 1834 to seek his fortune in the theater. He showed his play

Figure 8.1
Portrait of Claude Bernard.

Arthur de Bretagne to an illustrious critic of the day. The critic, learning of Bernard's previous job and apparently more impressed by his energy than by the play, suggested he try medicine instead of literature. (The critic and Bernard were later to be fellow "Immortals" in the French Academy.) Bernard was an indifferent medical student; nonetheless, he somehow fell into the hands and laboratory of François Magendie (1783–1855), professor of medicine at the College de France and head of one of the first laboratories devoted to experimental physiology.[2]

Magendie's father had been an active Republican in the French Revolution and, following Rousseau, had brought up his son as a free spirit. Magendie became a thoroughgoing materialist and was heavily influenced by the Ideologues, a group of revolutionary philosophers led by Pierre Cabanis (1757–1808) and A. L. C. Destutt de Tracy (1754–1836). They rejected Cartesian dualism, vigorously asserting that the mind was a "mere" function of the body and as Cabanis put it that "the brain was a bodily organ that . . . digests impressions and . . . secretes thought."[3]

Magendie had contempt for social convention and utter contempt for contemporary theories of medicine—indeed for the very idea of "theory" in science. For him, science meant only experiments and the facts that could be unambiguously and directly derived from them. He raised empiricism to a faith and denied that he was guided by hypotheses (as he obviously often was).

Before Magendie, much of physiology had been speculation and inference from anatomy and clinical medicine. Magendie established the importance of direct experiments on living mammals, usually cats, dogs, and rabbits. Even after their discovery in the 1840s, anesthetic agents were often not used in animal experiments, perhaps because of their depressing effect on nervous function: in this period experiments on the neural control of physiological function or on the nervous system itself were of central concern. In Magendie's (and Bernard's) time there was much less popular opposition to vivisection in France than in Great Britain; with the rise of a strong British antivivisection movement toward the end of the nineteenth century, this difference became even more pronounced.[4]

Perhaps Magendie's most famous discovery was of the law of spinal roots, also known as the Bell-Magendie law (i.e., that ventral spinal roots are motor and dorsal ones sensory). There was a long and bitter priority controversy with Charles Bell (1774–1842) over its discovery. In fact, Bell had originally proposed only the sensory functions of the dorsal roots; there is no reason to believe that Magendie knew of Bell's claims before he carried out and published his own experiments. Both halves of the law were physiologically demonstrated by Magendie, whereas the Englishman Bell (not a vivisectionist) had inferred the functions of the dorsal roots solely from anatomical observation.[5]

From Magendie, Bernard acquired a profound skepticism of established dogma and learned the techniques of vivisection that were the basis of the new animal physiology. He never practiced medicine and instead concentrated on research, eventually taking over Magendie's laboratory and chair. Bernard made a number of major experimental discoveries and theoretical advances that established him as the founder of modern physiology. Among his most important discoveries were the glycogenic function of the liver, the role of the pancreas in digestion, the regulation of temperature by vasomotor nerves (see box 8.1 and figure 8.2), the action of curare and carbon monoxide, and the vagal control of cardiac function. Most of this work was done early in his career, between 1843 and 1858, in a small damp cellar and with little funding.[6]

Although he continued some laboratory work for the rest of his life, Bernard became increasingly involved in two other concerns. The first was the political goal of establishing physiology, "experimental medicine," as an independent discipline. He was particularly concerned about separating it from clinical medicine, with its emphasis on intuition and "touch," and from chemistry, with its claims that the inorganic and organic could be treated equivalently. His second major interest was in broad theoretical issues such as the role of determinism in biology, the relation of theory and experimentation in biology, and the existence of phenomena common to both plants and animals and absent in the inorganic world. Among his new

Figure 8.2
Claude Bernard in his laboratory (Académie Nationale de Médicine, Paris). See box 8.1.

Box 8.1
Claude Bernard in His Laboratory

Figure 8.2 is an anonymous copy of a painting by L. A. L'hermitte made ten years after Bernard's death (Académie Nationale de Médicine, Paris). Bernard is the central figure, wearing an apron, and is surrounded by some of his most famous French students (Paul Bert, mentioned in the text, is standing third from the left). The setting is Bernard's laboratory at the College de France, which still can be visited today. The experiment illustrated is one in which the vasomotor functions of sympathetic nerves were demonstrated for the first time. Bernard is studying the effect of unilaterally cutting and stimulating the cut end of the cervical sympathetic nerve on the temperature of each side of the head of a rabbit. He discusses this experiment in *An Introduction to the Experimental Study of Medicine* (1865) to illustrate two principles of experimentation. The first was the importance of the choice of species, the rabbit being ideal here because the cervical sympathetic vascular nerves, unlike in other common laboratory animals, run separately from sensory and motor nerves.

The second was the value of hypotheses, even when wrong, as in this case.

On the basis of a prevailing theory and of earlier observations I had been led ... to make the hypothesis that the temperature should be reduced ... after severing the cervical sympathetic nerve in the neck.... The result was ... precisely the reverse of what my hypothesis, deduced from theory, had led me to expect; thereupon I did as I always do, that is to say, I at once abandoned theories and hypothesis, to observe and study the fact itself.... Today my experiments on the vascular and thermo-regulatory nerves have opened a new path for investigation and are the subject of numerous studies which, I hope, may some day yield really important results in physiology and pathology. This example ... proves that in experiments we may meet with results different from what theories and hypothesis lead us to expect.... This ... example ... gives us an important lesson, to wit: without the original guiding hypothesis, the experimental fact which contradicted it would never have been perceived.... Indeed I was not the first experimenter to cut this part of the cervical sympathetic in living animals.... But none of them noticed the local temperature phenomenon ... though this phenomenon must necessarily have occurred.... The hypothesis ... had prepared my mind [and my predecessors'] for seeing things in a certain direction.... We had the fact under our eyes and did not see it because it conveyed nothing to our mind. However, it could not be simpler to perceive, and since I described it, every physiologist without exception has noted and verified it with the greatest ease. (Bernard, 1961 [1865])

See also figure 8.3.

Figure 8.3

An unlabeled sketch from Bernard's *Cahier Rouge*, a diary in which he wrote (1850–1860) about not only things he had seen in the laboratory that day and the experiments or hypotheses that these had suggested, but also his constant search to express the experimental method in physiology, and to show how it related to other sciences (Bernard, 1967). The sketch seems to be a device for measuring the temperature in the two ears of a rabbit, presumably for an experiment like the one shown in figure 8.2.

and important theoretical concepts were those of internal secretions, reciprocal innervation, and, as we discuss in detail below, the internal milieu.[7]

Bernard was a consistent opponent of vitalism, arguing that biology never violated the laws of physics and chemistry. However, he did stress the emergent properties of complex biological systems much more than his German physiological contemporaries such as Helmholtz and du Bois-Reymond, who strove to reduce biological phenomena to physics and chemistry.[8]

The high point of Bernard's theoretical endeavors was the publication in 1865 of his *Introduction to the Experimental Study of Medicine*. It was an immediate success among scientists and physicians as well as philosophers and writers. Indeed, it remains in print to this day, even in English, and is still heralded as required reading for any prospective experimental biologist. One of its most timeless and attractive aspects is its autobiographical character; Bernard illustrates various principles and practices of experimentation almost exclusively from his own work. He does clean up the stories of some of his discoveries, however, omitting errors, blind alleys, and failed experiments.[9] Thus the book makes science seem easier than it really is.

MME. BERNARD AND MME. RAFFALOVICH

In 1845, near the beginning of his career, financial difficulties led Bernard into an arranged marriage with Fanny Martin, the daughter of a relatively well-off physician. Her dowry enabled him to avoid a rural practice and stay in research. The marriage was a disaster. Mme. Bernard bitterly resented her husband's low-paying research career and became an ardent antivivisectionist. Bernard's propensity to bring home opened-up and dying animals with various tubes stuck in them did not help matters. Finally, in 1869, when Bernard reached the peak of his career, they separated. Subsequently, she and her daughters founded a home for stray dogs and cats.[10]

After the separation, Claude Bernard became close to Marie Raffalovich, a Jewish intellectual from Odessa who was interested in science and

Box 8.2
From Bernard's Letters to Mme. Raffalovich

[1870, after a discussion of science, intuition, and superstition] The scientist, if he is to have great ability, must have imagination but he must master this imagination and coldly probe the unknown. However, if he lets himself be carried away by his imagination, he will be overcome by vertigo and, like Faust and others, fall into the chasm of magic and succumb to phantoms of the mind.

[1873, after a description of the history of the College de France] I follow in the tradition of my predecessors, who have all been men in the avant-garde of science, men of fighting spirit. I am fighting for physiology because it is the future of medicine. (Quotations from Bernard, 1978 [1869–1878])

philosophy. She attended his lectures, he visited her twice a week, and they often went to galleries and museums together. Unlike Bernard, she was an accomplished linguist and helped him with the foreign literature. Over the course of nine years he wrote over 500 letters to her, often when she was away on holiday with her family. Many of them have been published in two collections, and they yield a fine-grained account of his daily life and thoughts (see box 8.2). In 1876 she published a novel and he claimed that she was deserting him for the literary crowd. Then, in 1878, when she received news that he was very ill, she and her daughter went to nurse him in his final days. Mme. Raffalovich had her letters to Bernard destroyed after his death.[11]

HONORS AND FAME

Claude Bernard collected more honors and, arguably, became more famous than any French scientist before or after. He was elected to the Academy of Science, then the Academy of Medicine, and finally, most prestigious of all, he became one of the 40 "immortals" of the French Academy and eventually its president. He was commander of the Legion d'Honneur and a member of the Senate (a powerless front for the autocracy of Napoleon III). Bernard dutifully attended every Senate meeting but did not speak, even on such issues as academic freedom and rural medicine. When he died he was given the first state funeral ever afforded a scientist in France. Flaubert called it more beautiful and more stirring than the then-recent funeral of Pope Pius IX.[12]

From the height of his career until well after his death, Bernard was so famous that he became identified in the public mind as the stereotypical scientist, much like Albert Einstein in the twentieth century.[13] He appears in poetry, memoirs, and novels of the time, both in France and abroad (e.g., *The Brothers Karamazov*). Zola considered writing a novel in which a scientist is persecuted by his antivivisectionist wife, and wrote:

> I will make a scientist married to a backward bigoted woman, who will destroy his researches as he works. . . . I am tempted to model him after Claude Bernard, getting access to his papers and letters. It will be amusing . . .[14]

In the completed novel *Le docteur Pascal*, Zola moved away from Bernard as a model; the plot complications required a heredity researcher rather than a physiologist, but some similarities to Bernard remain. In his essay "The Experimental Novel," supposedly modeled after Bernard's experimental medicine, Zola manipulated plots and observed the behavior of his human characters just as Bernard manipulated physiological variables and observed their effects. "I have but one desire," Zola wrote. "Given a powerful man and an unsated woman, to cast them into a violent drama and scrupulously note down the sensation of these creatures." In fact this was hyperbole if not outright hype: Zola had begun his novel cycle before he was familiar with Bernard's writings (and before Bernard was famous).[15]

A bronze statue of Bernard engaged in vivisection was set up in front of the College de France after his death. The Germans melted it down in

World War II and it was replaced by a new statue in stone after the war.[16] This was destroyed during the student uprising of 1968 but has subsequently been replaced.

As Bernard had desired, his early play *Arthur de Bretagne* was published after his death. However, his widow and daughters claimed its preface defamed them and they successfully sued to have all copies destroyed. It had a radio production in 1936, and a second edition appeared in 1943.[17]

The Constancy of the Internal Environment

Bernard's ideas about the internal environment evolved from its first mention in 1854 until his death in 1878. He probably took the term from Charles Robin, a contemporary histologist who used *milieu de l'interieur* as a synonym for "the humors." Initially, for Bernard, the internal environment was simply the blood. But even at this stage, he understood that the temperature of the blood is actively regulated and that its constancy is particularly critical in higher animals. It was only later that he recognized that this constancy might be achieved through the vasomotor mechanisms he had discovered. At about the same time he realized that the glycogenic mechanism he had found controlled the constancy of blood sugar level. It was primarily on these two (limited) lines of evidence that he built his brilliant generalizations that unify the fundamental physiologies of the body:[18]

> The fixity of the milieu supposes a perfection of the organism such that the external variations are at each instant compensated for and equilibrated.... All of the vital mechanisms, however varied they may be, have always one goal, to maintain the uniformity of the conditions of life in the internal environment.... The stability of the internal environment is the condition for the free and independent life.[19]

These generalizations both summarized many of Claude Bernard's experimental achievements and provided a program for the next 100 years of general physiology. Although Bernard made these ideas central to his well-attended lectures and his widely disseminated writings, they were ignored in his lifetime and they had no impact at all until about 50 years later. Indeed, Bernard's ideas on the internal environment are hardly mentioned in the extensive 1899 biography by Michael Foster, the distinguished Cambridge physiologist; they are not mentioned at all in the twelve-page obituary in the American journal that had published much of Bernard's research or in a 1931 biographical essay by the eminent historian of science Henry Sigerist. Whereas the 1911 *Encyclopedia Britannica* is totally silent on the constancy of the internal environment, the 1975 edition calls it Bernard's "most seminal contribution."[20]

An exception to the nineteenth-century silence on Bernard's internal milieu was George Henry Lewes (1817–1878), the Darwinian publicist (and life partner to George Eliot, in which capacity he made the inside back cover of the *New Yorker* in 1998). In his *The Physical Basis of Mind*, Lewes used the concept of the internal environment to answer an objection to evolution by the American anti-Darwinian Alexander Aggasiz.[21] The latter had claimed that the diversity of animals in the same environment argued against the possibility of natural selection. Lewes countered by stressing the similarities in their internal environment. Bernard himself varied between skepticism and dismissal of Darwinism, reflecting his view that if biological phenomena were not experimentally demonstrable they were of little validity.[22] Yet, it was only when the profound evolutionary significance of the constitution of the internal environment was realized that Bernard's idea finally had a major impact on physiology.

The development that catalyzed the understanding of Bernard's milieu interieur was the comparison of the ionic concentrations of body fluids with those of seawater.[23] In 1882, Leon Fredericq observed that the body fluids of ocean crabs, lobsters, and octopuses were about as salty as seawater, whereas marine fish, like freshwater ones, were much less salty. (He made these observations initially by taste.) He realized that this was the first evidence for Bernard's idea that the internal milieu becomes increasingly independent of the external environment as one ascends the "living scale," thereby providing the basis for the "free life" of higher organisms.[24] Fredericq had studied in Paris with Paul Bert, a major student, collaborator, and biographer of Bernard (shown in figure 8.2). In marked contrast to Bernard, however, Fredericq interpreted his comparative observations as evidence for the evolution of the independence of the internal environment from the external one. By the end of the century, evolutionary thinking had finally made the constituents of the internal environment a meaningful subject. Independently, René Quinton and Archibald Macallum took the next step, arguing that life arose in the sea and that body fluids represented the original seawater that had been enclosed within the skin. More generally, it became clear that a major trend in evolution was the development of increasingly sophisticated mechanisms whereby the internal environment is protected from the external world.[25]

In the first decades of the twentieth century, Bernard's ideas about the importance of the internal environment entered the mainstream of mammalian physiology both as a central explanatory concept and a program for research. Among the major British figures explicitly relating their work closely to Bernard's idea were William Bayliss and E. H. Starling, codiscoverers of secretin, the first hormone identified; J. S. Haldane (J. B. S. Haldane's father) and Joseph Barcroft, pioneers in the regulatory functions of breathing; and C. S. Sherrington, a founder of modern neurophysiology. Starling seconded Macallum and Quinton's ideas on the evolution of the internal environment and later coined the term "the wisdom of the body" for the maintenance of the internal constancies that Bernard had postulated.[26] Barcroft claimed that the "principles...of the fixity of the internal environment have been as thoroughly established as any."[27] Haldane noted that Bernard's conception "sums up and predicts" his own work on the regulation of blood composition by respiration.[28] Sherrington suggested that "the nervous system is the highest expression of...the milieu interieur."[29]

In the United States, the chief advocates of Bernard's constancy ideas were L. J. Henderson and Walter B. Cannon, longtime members of the Harvard Medical School faculty. Henderson related his work on the maintenance of blood pH directly to Macallum's marine biology as well as to Bernard.[30] He helped bring Bernard to a wider American audience both in his introduction to the American translation of Bernard's *Introduction* and in his own influential book, *The Fitness of the Environment*.[31]

Walter B. Cannon was particularly instrumental in making Bernard's ideas central to the neurophysiology and psychology of the time. He coined the term *homeostasis* for the tendency of the mammalian organism to maintain a constant internal environment.[32] His own major discoveries were in elucidating the role of the sympathetic nervous system in maintaining homeostasis; he brought these to the educated public in the classic *The Wisdom of the Body* (1932). Cannon viewed behavior as a homeostatic mechanism: shivering, seeking shelter, and putting on a coat were all examples of homeostatic mechanisms of temperature regulation. Writing at the height of the Great Depression, he suggested that some institutional arrangements for social homeostasis were sorely needed:

> The main service of social homeostasis would be to support bodily homeostasis. It would therefore release the highest activities of the nervous system for adventure and achievement. With essential needs assured, the priceless unessentials could be freely sought.[33]

J. B. Watson and other early behaviorists such as Curt Richter rejected the myriad of previously postulated central drives as explanations for motivation. They turned instead to the experiments of Cannon for alternative and peripheral mechanisms of motivation and considered "motivated" behavior as a homeostatic mechanism. Thus, following him, they viewed thirst as a result of dryness in the mouth, which, when signaled to the brain, elicited drinking. Similarly, hunger was caused by stomach contractions ("pangs"), which signaled the brain to elicit eating. Extrapolating beyond Cannon, they interpreted sexual motivation to be due to tension in the gonads.[34]

Both Cannon and Henderson had extended Bernard's ideas of self-regulation from the realm of bodily fluids to the wider social environment.[35] The idea of self-regulation was extended even further to include the nonbiological world by Arturo Rosenblueth (one of Cannon's collaborators), Norbert Weiner, and J. Bigelow.[36] In the context of World War II control and communication systems, they pointed out that negative feedback covered self-regulation both in the nervous system and in nonliving machines. Soon after, Weiner coined the term *cybernetics* for "the entire field of control and communication theory, whether in the machine or in the animal."[37] Today, cybernetics, a formalization of Bernard's constancy hypothesis, is viewed as one of critical antecedents of contemporary cognitive science.[38]

Why Was the "Constancy of the Internal Environment" Not Understood in Bernard's Time?

Despite the emphasis with which he repeatedly promulgated it, Claude Bernard's insight that the "constancy of the internal environment is the condition for the free life" had no significance (indeed, no meaning) for biologists for more than 50 years. There seem to have been several reasons for this inability to process his idea. One was that Pasteur's new bacteriology and its omnipresent, omnipotent germs were dominating the biomedical zeitgeist. Another, as discussed above, was the gap between evolutionary thought and general physiology. When this gap began to be closed through the comparison of the constituents of seawater and the bodily fluids at different phylogenetic stages, the constancy of the internal environment suddenly took on new and accessible meaning. Finally, the tools, techniques, and concepts for adequately measuring the internal environment were simply not available in Bernard's time and for the rest of the century. For example, the work of Haldane, Henderson, and Barcroft required the development of organic and especially physical chemistry, as well as techniques for measuring ions, gases, and other components of the internal environment; the work of Sherrington and Cannon required the replacement of the reticular doctrine by the neuron doctrine, and the development of the cathode-ray tube oscilloscope and electrical stimulating devices.[39]

In the history of biology there have been those, such as Gregor Mendel and Emmanuel Swedenborg, who were so far ahead of their time that they died unrecognized for their scientific work.[40] Claude Bernard, by contrast, received every possible recognition as a scientist, yet what is today considered his most salient contribution had to wait half a century for advances in theory and practice to make it meaningful.

NOTES

This chapter was originally published as an article with the same title in *The Neuroscientist* (4: 380–385 [1998]). Parts of it were included in my later paper "Three before their time: Neuroscientists whose ideas were ignored by their contemporaries," published in *Experimental Brain Research* (192: 321–334 [2009]).

1. Sigerist, 1933.

2. Olmsted, 1939; Olmsted and Olmsted, 1952; Grmek, 1970a.

3. Olmsted, 1944; Grmek, 1970b; Temkin, 1946a.

4. Elliott, 1987; Manuel, 1987; Rupke, 1987; Schiller, 1967; French, 1975.

5. Cranefield, 1974.

6. Olmsted, 1939; Olmsted and Olmsted, 1952; Grmek, 1970a; Grande, 1967; Robin, 1979.

7. Grmek, 1970a; Coleman, 1985; Wasserstein, 1996.

8. Bernard, 1961, 1974; Temkin, 1946b.

9. Grmek, 1970a; Holmes, 1974.

10. Olmsted, 1939; Olmsted and Olmsted, 1952.

11. Olmsted, 1939; Olmsted and Olmsted, 1952; Virtanen, 1960; Bernard, 1950, 1978.

12. Olmsted, 1939; Olmsted and Olmsted, 1952; Grmek, 1970a.

13. Olmsted, 1939; Olmsted and Olmsted, 1952; Virtanen, 1960.

14. Virtanen, 1960.

15. Virtanen, 1960.

16. Olmsted and Olmsted, 1952.

17. Olmsted and Olmsted, 1952.

18. Holmes, 1963, 67; Langley, 1973.

19. Bernard, 1974.

20. Olmsted, 1967; Foster, 1899; Flint, 1878; Holmes, 1965.

21. Lewis, 1877.

22. Bernard, 1961, 1974; Virtanen, 1960; Fredericq, 1973; Petit, 1987.

23. Holmes, 1965.

24. Holmes, 1965; Fredericq, 1973.

25. Holmes, 1965; Macallum, 1926.

26. Starling, 1909.

27. Barcroft, 1932.

28. Haldane, 1931.

29. Sherringon, 1961.

30. Henderson, 1928.

31. Henderson, 1958, 1961

32. Cannon, 1929.

33. Cannon, 1963.

34. Watson, 1924; Richter, 1927; Cannon, 1963.

35. Cannon, 1963; Henderson, 1935.

36. Rosenblueth et al., 1943.

37. Wiener, 1961.

38. Gardner, 1985.

39. Virtanen, 1960; Olmsted, 1967; Petit, 1987.

40. Gross, 1997a.

9

Bartolomeo Panizza and the Visual Brain

with Michael Colombo and Arnaldo Colombo

Bartolomeo Panizza (1785–1867) was the first person to produce experimental and clinicopathological evidence for a visual area in the posterior cerebral cortex. This was, arguably, the first systematic evidence for the localization of function in the cerebral cortex. We here provide the first translation of this work entitled "Observations on the optic nerve," originally published in Italian in 1855. Published before Broca's and Fritsch and Hitzig's work,[1] which are usually considered to have initiated cerebral localization, Panizza's discovery of visual cortex was ignored until after its independent rediscovery. It was then largely forgotten. In this article we briefly review the knowledge of the brain in Panizza's time, summarize his scientific career, consider why his paper on visual cortex was lost, and then provide the first full translation of this paper originally published in 1855.[2]

The Cerebral Cortex in 1855

The first half of the nineteenth century was a period of conflicting views on the functions of the cerebral cortex.[3] In the eighteenth century, the standard view of the brain was that of Albrecht von Haller (1708–1777), a Swiss naturalist, anatomist, and physiologist and the dominant figure in brain anatomy and physiology. He believed that all parts of the brain had the same organization and functioned in the same way. This unity of the brain, he thought, reflected the unity of the soul. This view was challenged at the beginning of the nineteenth century by Franz Joseph Gall and Johann Gaspard Spurzheim: their phrenological system postulated that the cerebral cortex was a set of organs with different psychological functions (see chapter 4 and figures 4.5 and 4.6). These 27 organs were concerned with "affective" or "intellectual" faculties; basic sensory and motor functions were thought to be subcortical, residing in the thalamus and corpus striatum, respectively.[4]

Gall's theories of punctate localization in the cortex were attacked by Pierre Flourens (1794–1867). He reported that lesions of the cerebral cortex had devastating effects on willing, judging, remembering, and perceiving, but that the site of the cortical lesion did not matter. However, lesions to other structures such as the cerebellum and the medulla did produce different symptoms. Flourens's findings, although a refutation of Gall's methods and specific localizations, were actually a confirmation of Gall's general attempt to localize functions in different parts of the brain and of his emphasis on the higher roles of the cerebral cortex.[5]

Although Gall and Spurzheim's use of cranial morphology ("bumps") was soon rejected by the scientific community, Gall's ideas of punctate localization spurred the search for different cortical organs. For example, Jean-Baptiste Bouillaud (1796–1881) tried to support some of Gall's localizations, such as of language, by direct clinicopathological examination of human patients. Two major discoveries finally established the idea of localization of function in the cerebral cortex. The first was Broca's report of a relationship between damage to the left frontal lobe and deficits in speaking.[6] The second was Fritsch and Hitzig's demonstration of specific movements from electrical stimulation of specific regions of the cortex.[7] At the time both these discoveries were viewed as vindication of Gall's ideas of the localization of function in the cortex.[8] Both occurred after Panizza's article, here translated, was published. (See chapter 4.)

WHO WAS BARTOLOMEO PANIZZA?

Panizza was born in Vicenza, Italy, and took his degree in surgery at the University of Padua in 1806 and in medicine at the University of Pavia in 1810. With the help of his father, a distinguished physician, he chose an academic research career rather than going into medical practice. Panizza became professor of anatomy at Pavia at the early age of 32 and held this position for the next 50 years. He appears to have "enjoyed the very highest reputation among Italian and foreign scientists"[9] and received a number of awards and honors for his research as well as appointments as dean of the medical faculty and rector of the University of Pavia. Panizza worked on a variety of subjects including the lymphatic and circulatory systems, the parotid gland, and the cranial nerves. His last major paper was the one here translated, published when he was 70.[10]

THE ACHIEVEMENTS OF "OBSERVATIONS ON THE OPTIC NERVE"

This study utilized a wide range of species from fish to humans, and two main techniques. The first technique was to unilaterally blind animals (enucleation) and then trace the resultant degeneration. This method, known as the "atrophic degeneration method," was later rediscovered and credited to von Gudden (1824–1889).[11] Panizza found it worked particularly well when the enucleation was done in infancy and the anatomy in adulthood. Using this method he inferred that, in mammals, parts of the thalamus and the posterior cerebral cortex were visual in function. The second principal method that Panizza used was to make lesions and observe the resultant behavior. When he made lesions in the brain regions in which he had found degeneration after enucleation, he found the animals to be blind contralateral to the lesion. Thus, in mammals, after lesions of the thalamus or the posterior cortex he observed contralateral blindness, confirming his idea that these were visual structures (and that the decussation of the optic pathways was total in mammals as well as other animals). He further supported this interpretation by observations of pathology in the thalamus and posterior cortex in two human patients who, he thought, were blind contralateral to the lesion. It is very likely that Panizza interpreted his results as supporting Gall's view that "the brain [is] a complex of organs" but added that "it is not easy to determine their number, the seat of each one, and their respective functions."

WHY WAS PANIZZA'S PAPER ON THE VISUAL SYSTEM IGNORED?

Panizza's work on the visual areas of the brain does not seem to have been cited in the scientific literature until after the report of Munk in 1878 of a visual area in the occipital cortex of the dog. Luciani and Tamburini repeated Munk's observations,[12] and Tamburini subsequently cited Panizza's work for the first time.[13] Subsequently, Panizza's work has usually been ignored in discussion of the history of localization of function, although both Finger and Polyak mention him, and there has been a recent surge of interest in him by his countrymen.[14] Why was Panizza's paper ignored at the time, particularly since he was an established scientist and the subject of localization of function was one of much interest at the time? One reason may have been because it was in Italian and in a local publication. However, such publications were normally exchanged with the Royal Society and other scientific societies. Another reason may have been because at this time the general view, in fact the view of both Gall and Flourens, was that the cerebral cortex was devoted to higher "psychic" functions and subcortical regions were the highest centers for vision and other senses. Panizza's work seems yet another example of a work being too "ahead of its time" to be grasped by its contemporaries.

ON THE TRANSLATION

"Observations on the optic nerve" was originally delivered as a speech on April 19, 1855, to the Lombardy Institute of Science,[15] and published later that year.[16] Throughout the manuscript we have placed the modern-day names of anatomical structures in brackets.

Translation of Panizza's 1855 "Observations on the Optic Nerve"

The object of science is truth, of art beauty

—*Giordani*

Although many famous anatomists have conducted serious and important studies on the origins of the optic nerve, and have ascertained, especially with the assistance of embryology and live dissection, that the principle origin of this nerve is the bigeminal eminence[17] [optic tectum in fish or superior colliculus in mammals], they are no longer in agreement over the parts that are contributed by the optic thalamus, the cerebral peduncles, the tuber cinereum, the lateral walls of the infundibulum of the third ventricle, etc. For the purpose of clarifying the true relationship of the optic nerve with the brain I believe that it was worthwhile for me to also look into this.

In fish (a class of animal in which the brain is very different from one genus to the other) I examined only the brains of river fish, especially that of *Exos lucius* [pike] and *Cyprinus tinca* [tench], of which I wanted to determine the external and internal parts.

In pike the tubercles or anterior lobes [telencephalon], that some call the cerebral hemispheres (Table VIII [reproduced here in figure 9.1], Figure 1b), of conical form, knobby, and ash-colored are divided in two masses by a barely visible almost transverse but deep sulcus; the small anterior mass that forms the apex of the cone belongs to the olfactory nerve; the other bigger mass covers a part of the first one and is on top of it and attached with a part of its substance at the root of the olfactory nerve, so that it is part of the same and joins the hollow lobe[18] [optic tectum] located posteriorly. The

---

205

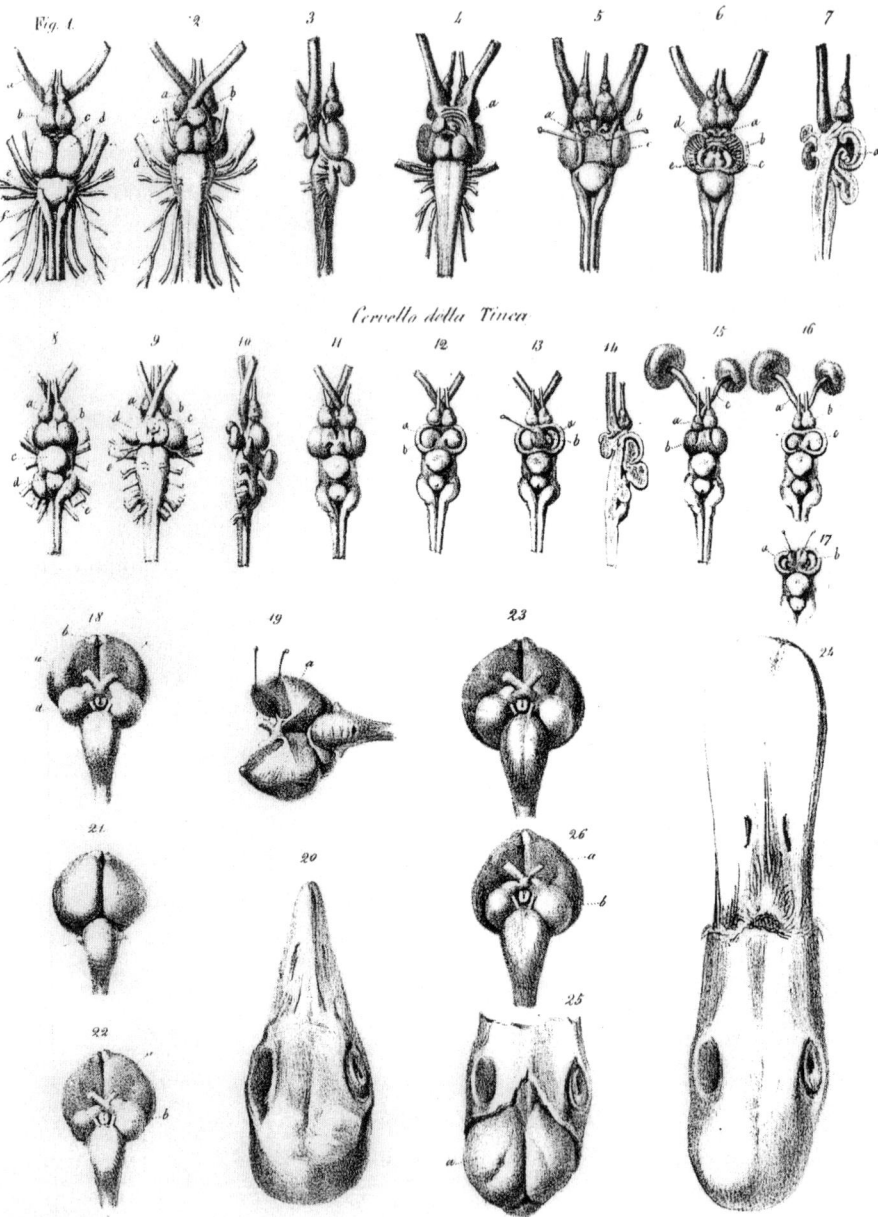

Figure 9.1

Panizza's Table VIII. The figures on the first line represent the brain of the pike shown in various positions and sections. Figure 1. Superior surface of the brain: (a) optic nerve, (b) anterior lobe, (c) small tubercle, (d) hollow lobe, and (e) small brain. Figure 2. Inferior surface of the brain: (a) decussation of the optic nerves, (b) pituitary body, (c) inferior lobes of the brain, and (d) elongated medulla. Figure 3. Side view of the brain. Figure 4. Inferior surface of the brain after removal of the superimposed optic nerves: (a) arcuate columns. Figure 5. Superior surface of the brain with removal of the hollow lobes to show the commissure, the superior internal upper roots of optic nerves, and the two small tubercles: (a) commissures, and (b) the meeting of two nerve fascia which comprise the optic nerve or hollow lobe. Figure 6. Shows the two small tubercles and the internal parts of hollow lobes: (a) small tubercles, (b) walls of the hollow lobe, (c) corpus striatum, (d) radiating lamina, and (e) quadrigeminal eminence. Figure 7. Side view of the vertical section of the brain: (a) shows that the quadrigeminal eminence are just the refoldings of the walls of the hollow lobes. Brain of tench (Figures 8–14 represent the brain of the tench in various positions.) Figure 8. Superior surface of the brain: (a) anterior lobes, (b) hollow lobes, (c) small brain, (d) appendix of the small brain, and (e) leg of the small brain. Figure 9. Brain shown from the inferior surface: (a) crossing of optic nerves, (b) pituitary body, (c) inferior and reniform lobe, (d) hollow lobe, and (e) elongated medulla. Figure 10. Side view of the brain. Figure 11. Shifting of the two hollow lobes to show the upper roots of the optic nerves and the fibrous apparatus positioned between them. Figure 12. Section of the hollow lobes and the objects contained therein: (a) contour of the section, and (b) internal eminence that almost entirely fills the cavity. Figure 13. Shows the parts of the hollow lobe located under the internal eminence: (a) corpus striatum, and (b) small body that is part of the optic nerve. Figure 14. Side view of the vertical section of brain. Figure 15. Brain of a tench that was blinded in the right eye when it was young, and dissected after 1 year: (a) atrophy of left anterior lobe, (b) atrophy of hollow lobe, and (c) atrophy of the nerve of the right eye. Figure 16. Another similar example: (a) developed eye and optic nerve, (b) developed right anterior lobe, and (c) hollow lobe and pronounced objects contained therein, while there is atrophy of the right eye, its nerve, and left anterior lobe and hollow lobe. Figure 17. Another example of a tench that when young was blinded in one eye, showing the alterations that occurred in the objects contained in the hollow lobe: (a) atrophy of the hollow lobe in which the internal objects were also atrophied, and (b) well-developed right hollow lobe containing the corpus striatum, the internal lobe, and the other small tubercle likewise developed. Figure 18. Brain of a chicken: (a) cerebral hemisphere, (b) olfactory lobe, (c) optic nerve, and (d) optic lobe. Figure 19. Removal of the two cerebral hemispheres to demonstrate the radiating lamina of the flat surface of the cerebral hemispheres: (a) the neural peduncle of the radiating lamina that joins the optic nerve before arriving at the aja, the other part turns inside the foot of the brain and terminates at the anterior apex of the optic thalamus. Figure 20. Head of a chicken blinded in the left eye as a chick; after 8 months the skull was opened and an increase in the left side and a depression on the right side was found. Figure 21. Brain of the same chicken; the left brain hemisphere is larger and protruding on the posterior superior part. Figure 22. The same shown from the inferior surface: (a) the left hemisphere is more developed, and (b) the optic lobe is bigger, as is the nerve, while the opposite optic lobe and nerve are atrophied. Figure 23. Brain of a duck. Figure 24. Head of a duck that was blinded in the left eye a few days after birth; after a few months the eye was atrophied, the orbit was smaller. The right eye is well developed, the skull in the left superior posterior part is larger, and on the right side it is depressed. Figure 25: (a) More developed left hemisphere. Figure 26. Inferior surface of the same brain that shows a larger left hemisphere: (a) of the optic lobe and corresponding nerve, and (b) atrophy of the right optic lobe and nerve.

hollow lobes (Figure 1d) are big, oval, and tinted ash-white; between their anterior extremities and the anterior lobes some anatomists have noted a small ash-colored mass that they call the pineal body. In my examination of various pike brains I did not see this body; instead I noticed, pulling back a bit the anterior extremities of the hollow lobes, two ash-colored small tubercles (Figure 1c), at times just visible, one to the right, the other to the left, that became white, and converging unite at the medial line (Table VIII, Figure 6a). When the hollow lobes are opened, we can see that their walls are comprised of three layers, an external fibrous-white layer, a middle ash-colored layer, and an interior layer that is medullary and fibrous.[19] In the inferior part of the interior layer there are some fibrous fascia, white, composed of many slender fibers (Figure 6d) that disperse like a fan on the sides and roof of the cavity. These lamina radiate anteriorly, folding with some fibers which, close to the medial line, join with those from the other lobe, thus forming a medullary commissure. The ash-colored body on top of the base of the radiating fibers, and called the corpus striatum or optic thalamus[20] (Figure 6c), does not converge to form the aforementioned fibers. In the common cavity of the hollow lobes at the medial posterior part one can discern a ridge formed by four eminences, similar in their disposition and configuration to the quadrigeminal eminence[21] (Figure 6e), but they are only bends of the ash-colored medullary lamina of the hollow lobes (Figure 7a). The small brain [cerebellum] that is situated posteriorly of the aforementioned hollow lobes, has a conical form, and with its margins is united at the lateral eminence of the elongated medulla [medulla oblongata]. The substance of the small brain is very reddish, is rich in vessels, and has at its core a fibrous mass. On the external surface inferior to the brain are two lobes called the inferior lobes, oval-shaped or kidney-shaped, that some consider the mammillary eminence (Figure 2c). These lobes, however, located beside the inferior branch of the optic nerve, do not supply fibers to it.

For the nerves I examined only the optic nerves in their passage from the eye to the brain. When one nerve meets the other it passes over it;

sometimes the nerve on the right crosses over the nerve on the left; or sometimes the opposite happens, however, they never join, remaining attached only by a cellular fabric. Behind this intersection they are joined by the fibers from the tuber cinereum, at which point each nerve is divided into two fascia, a superior slender one that runs back along the superior internal side of the hollow lobe, the other, thicker, that wraps along the inferior internal side of this same lobe. Between the two inferior fibrous fascia before the peduncle of the pituitary gland, one finds the medullary lamina of Haller, that joins these two fascia, and behind this are two medullary fibrous arches one anterior, the other posterior, that convexes anteriorly (Figure 4a): these arches are connected to the interior fibers of the inferior optic fascia.

The brain of the *Cyprinus tinca*, both in its external contour and in its internal parts, differs significantly from the previous description. The anterior lobes, referred to by some as the cerebral hemispheres, are more conical in shape; the hollow lobes have a rounder shape than those of the pike. The two lateral eminences of the elongated medulla are more arched and protruding; furthermore, behind the small brain, along the medial line of the fourth ventricle, there is a very notable cerebral appendix. The hollow lobes, unlike those of *Exos lucius*, internally have an eminence that almost fills the cavity (Table VIII, Figure 12b); the eminence is unattached in all its contours except at its inferior part where it is attached to the floor of the cavity. Removal of this eminence reveals another small eminence, referred to by some as the optic thalamus or corpus striatum (Figure 13a), under which one can find another very small projecting body that is directly connected to the inferior branch of the optic nerve (Table VIII, Figure 13b).

The relationship of the optic nerve of the tench to the hollow lobes and other structures is no different from that of the pike.

In order to better understand the relationship of the optic nerve with other brain structures, I decided to partially blind young pike and tench, leaving them afterwards in a vast fishery for many months and up to 1 year or more. The following alterations occurred (which are shown in Figures 15–17; Table VIII): atrophy of the optic nerve of the blind eye, the decrease in volume of the opposite anterior lobe, especially that of the hollow lobe, in which the walls were atrophied; and in the brain of the tench were atrophied any object contained therein. Meanwhile in the other hollow lobe we found above normal swelling of the walls and the eminences within, and in particular the papilla above this, that is, the corpus striatum or optic thalamus (Table VIII, Figure 17b).

From this I think we can infer that [the] optic nerve of fish is derived little from the anterior lobes, more from the walls of the hollow lobe, but not from the objects contained therein, and in particular the corpus striatum or optic thalamus, and none from the eminences situated externally and lower to the inferior lobe, because between this and the optic nerve there is no communication, and because there was no evidence of any alteration in the pathological cases.

In birds, although the primary origin of the optic nerve is the optic lobe [optic tectum], it is certain that the optic thalamus also contributes to its formation, as well as the peduncle of the brain with two or three fibers, the tuber cinereum, and the radiating lamina [optic radiations]. These (Table VIII, Figure 19a) descend along the flat surface of the cerebral hemisphere and concentrate in a nerve fascia from which a part enters the optic nerve before it joins into the chiasm; the other part, directed to the back circles the leg of the brain to then end in the external extremity of the optic thalamus, that in addition to being connected to the optic nerve, also finds itself in relation with the medullary fascia that proceeds from the cerebral hemisphere.

To clarify the nature and importance of the relationship of the optic nerve with the various structures previously discussed, I conducted several experiments. Of birds I selected the crow, because in addition to being a lively and strong animal, it has a soft skull that can be cut with a simple knife, thus exposing the brain without damaging any of its functions. Exposing a cerebral hemisphere, I raised with a probe the posterior and lateral part, and with a very thin scalpel shaped like a spear I touched and damaged the optic lobe, once in a transverse direction, and once from front to back. Every time that I made any soft touch the animal did not react, but when I tried to penetrate the substance with a needle, not only did the senses appear, but there were even convulsive movements, which I will discuss at another time. With respect to sight, I have verified that which the celebrated Flourens and others have observed, that is, a loss of vision from the opposite eye not only after a profound wound, but also from a small lesion, hence the integrity of the optic lobe is absolutely necessary for the exercising of this function. In fact, when the crow that sustained the above injury was allowed to wander, it walked as quickly as before, but ran at every step into the wall and other objects placed on the side of the damaged eye: a finger or other object placed near this eye was not seen, although the iris reacted normally. Kept alive for 2 days, the bird ate and maintained its usual lively status and quick readiness of movements, even though the brain hemisphere was exposed: the sight from the damaged eye remained constantly abolished. After bleeding the animal, we found on the right optic lobe a linear injury in the middle part of its superior surface that was deep two-thirds the length of the line, with small traces of blood on top of the damaged region. In other crows, using the same cautious procedure, I damaged instead the posterior part of the right optic thalamus above the optic lobe: the same result was obtained, that is, the loss of sight from the opposite eye, persistence of movements of the iris, normal movements of the entire body; in the sections we found the optic thalamus just barely damaged more or less next to the optic lobe.

In other crows I cut longitudinally into the ash-colored layer of the cerebral hemisphere close to the superior margin of the flat surface, and in doing so also involved the ash-colored mass that carries the fibers of the radiating lamina. The crows remained blind in the opposite eye, movements of the iris were maintained, as were the senses and movements of the body. The anatomical examination conducted after 1 or 2 days confirmed that the ash-colored layer from which originates the radiating lamina was affected by the cut for two-thirds of its length. Following up, I had a great desire to know what alterations would happen to the sight after transverse sections of the cerebral hemisphere at various distances from its anterior extremity. For that purpose, after assuring myself of the relationship of various points of the skull with the parts contained therein, I made, here and there, small holes in the skull, into which I inserted a slender spear-shaped scalpel, thus cutting in transverse a small portion of the anterior extremity of the cerebral hemisphere the size of two grains of rice put together: the animal showed no signs of suffering; when placed on the ground the animal was free and lively in its movements as if nothing had happened to it; however, the sight from the eye opposite to the damaged cerebral hemisphere was lost, but the pupils continued to move. Killed a day later, I ascertained the precision of the incision that I made. The same result was obtained with a not-too-deep transverse section either one-third, one-half, or one-fifth of the posterior section of the cerebral hemisphere. And even after removal of a small portion from the posterior margin of the cerebral hemisphere where it is pointed, the crow lost sight in the opposite eye. I would also add that the same lesion made to both cerebral hemispheres caused complete blindness without affecting other vital functions.

In other crows, I removed the right half of the dome of the skull, then cut the meninges, lifted the posterior cerebral hemisphere, and with two hits of the scissors I removed all of it: I immediately observed paralysis of the left side, but after a short time the animals recovered their movements, but the left eye remained blind yet the pupils still moved: the other eye saw perfectly. For the 2 or 3 days that they were left to live following the operation they were lively, they ate and walked normally. The section demonstrated that the hemisphere removed was that overlying the corresponding optic thalamus. I also decided to cut in transverse the anterior eighth part of each of the cerebral hemispheres: this was followed by a complete loss of sight, but the pupils remained mobile. The crows were doing nothing other than jumping, hitting everything they encountered. When I removed both hemispheres overlying the optic thalamus, the animals hemorrhaged heavily: every time I managed to stop the hemorrhage with pulverized resin, I noticed that after a short while the crows were able to stand up but remained immobile in that same position: if I shook them they moved automatically, and I was able to verify that their sight and hearing was totally lost. In fact, I was unable to confirm the observations made by the celebrated Flourens and Longet that a large amount of sight was spared, such that birds deprived of both brain hemispheres moved their head in the direction of the light that was presented: this I could never verify. In these experiments of removal of the cerebral hemispheres, I had time to observe that some movements are still possible in the animals on which these procedures were done, movements that at first appear to be voluntary, but in fact are not: when I opened the beak of one of these and introduced a piece of meat I saw it shaking its head as if it wanted to remove the meat from its mouth. By close observation I was able to determine that this movement was caused by the reflex action of the respiratory nerves, since this does not happen if the piece of meat introduced into the mouth does not come to rest against the internal aperture of the nostrils or the airway aperture, thus blocking the free intake of air for respiration.

In other experiments I blinded one eye of various baby chickens or ducks just born or a few days after birth. I kept them alive for many months, some even 1 year, and found that in all of them the optic nerve was atrophied, more or less ash-colored and gelatinous; a state also present in the square aja[22] that divided into two ash-colored fascia hugging the opposite nerve; in this manner the right nerve passes in between [the] left one; and if I follow this ash-colored fascia over the square aja it seems the crossing was total. In others I observed that the part that was most atrophied was the optic lobe, of which the surface was completely ash-colored: in any case I saw only a small reduction of the optic lobe while the optic thalamus and the rest of the nerve were reduced quite a bit. With respect to the good eye, the corresponding optic lobe was well developed but the rest of the nerve was not; but there were no cases in which I found very enlarged the section of the nerve between the thalamus and the eye, and nothing

in the optic lobe. From all these anomalies, we can infer the various connections the aforementioned nerve has with the above-mentioned encephalon, relations that are in harmony with anatomical and physiological
observations.

In mammals the relationship of the optic nerve with parts of the brain
are more or less the same. The origins of the nerve are from the bigeminal
eminence, especially the nates[23] [superior colliculi], optic thalamus and its
appendices: the medullary fascia coming from the circumvolutions of the
posterior part of the cerebral hemisphere, some fibers from the cerebral
peduncles, from the lateral walls of the infundibulum of the third ventricle,
and from the tuber cinereum. These relationships are the same in the rabbit,
ox, lamb, horse, dog, and human: in the horse more than any other animal
it has been demonstrated how much the optic thalamus is the same, and
how its fibrous fascia, that exit and form the optic nerve, follow the fibrous
fascia that exit from the external side of the same thalamus, that go on to
then form the fibrous apparatus of the cerebral superior and posterior inferior circumvolutions.

Although from these anatomical data we can conclude that the above-
mentioned cerebral parts with which the optic nerves are in relation must
affect the integrity of the function of the nerve, I wished to also conduct
experiments with live sections. In conducting these experiments I took pre-
caution while damaging one or the other part of the brain, to make as little
damage as possible to the other objects. Therefore, making a very small
circle of holes, I trephined the skull of rabbits and especially dogs. I introduced a slender pointed knife as wide as a line,[24] and cut transversely
one-fifth of the anterior part of the brain hemispheres. The animal, let
alone, walked and ran as if nothing had hurt it: its muscular energy and its
intellectual faculties did not appear diminished, but the sight of the opposite
eye was damaged. In other animals, cutting transversely the thickness of the
corpus striatum, I noted always loss of sight from the opposite eye and
nothing more. In some, when the anterior part of the optic thalamus was
cut, the opposite eye was blind without the animal suffering any other
damage. Finally, in a dog, I uncovered one section of the brain, very much

below the parietal hump, and removed a small portion of substance: nothing happened, except blindness in the opposite eye. The same operation conducted simultaneously in both cerebral hemispheres resulted in complete blindness. These experiments replicated several times prove to me that truly the parts of the brain that are in anatomical relation with the nerve important for sight exercise an important influence over them.

To further confirm my assumption I will succinctly explain the pathological facts observed in rabbits, dogs, ox, horses, and donkeys.

In several rabbits just a few days old I removed the corner of the left eye, emptying it thus of its humor. After 1 year we checked the resulting effects. The left eye, and also the orbit (Table IX [here shown in figure 9.2], Figure 1), were atrophied, while the right eye and orbit were overdeveloped. Exposing the skull, the posterior and superior part of the left side was more protruding, and this I also observed in the corresponding cerebral hemisphere. I also found a significant difference between the two optic nerves; the left was reduced and very slender, the right was quite well developed in the back of the square aja (Table IX, Figure 2) and circled the cerebral peduncle and flattening out it enlarged over the left optic thalamus, over the geniculate eminences and on the superior surface of quadrigeminal eminence (Table IX, Figure 1e–h)

Blinding one eye in several pre-weaned dogs, and 1 year later observing the superior part of the skull opposite to the good eye 1 year later, revealed a greater rise (Table IX, Figure 3a), a rise that is also observed in the corresponding cerebral hemisphere. Cutting the cerebral hemispheres at the level of the corpus callosum, and removing them in order to reveal the vault with three columns, I saw its margin, that was ugly, very large, and rounded, such that one might consider it the continuation of Ammon's Horn, more elevated along the left side than on the right; by also removing the vault, I realized that this difference depended on the major development of the objects contained in the left lateral ventricle compared with that on the right. Between these objects the more enlarged were clearly the optic thalamus with its geniculate body, and the corpus striatum (Table IX, Figures 4a–c). Between the superior quadrigeminal eminence there did not

---

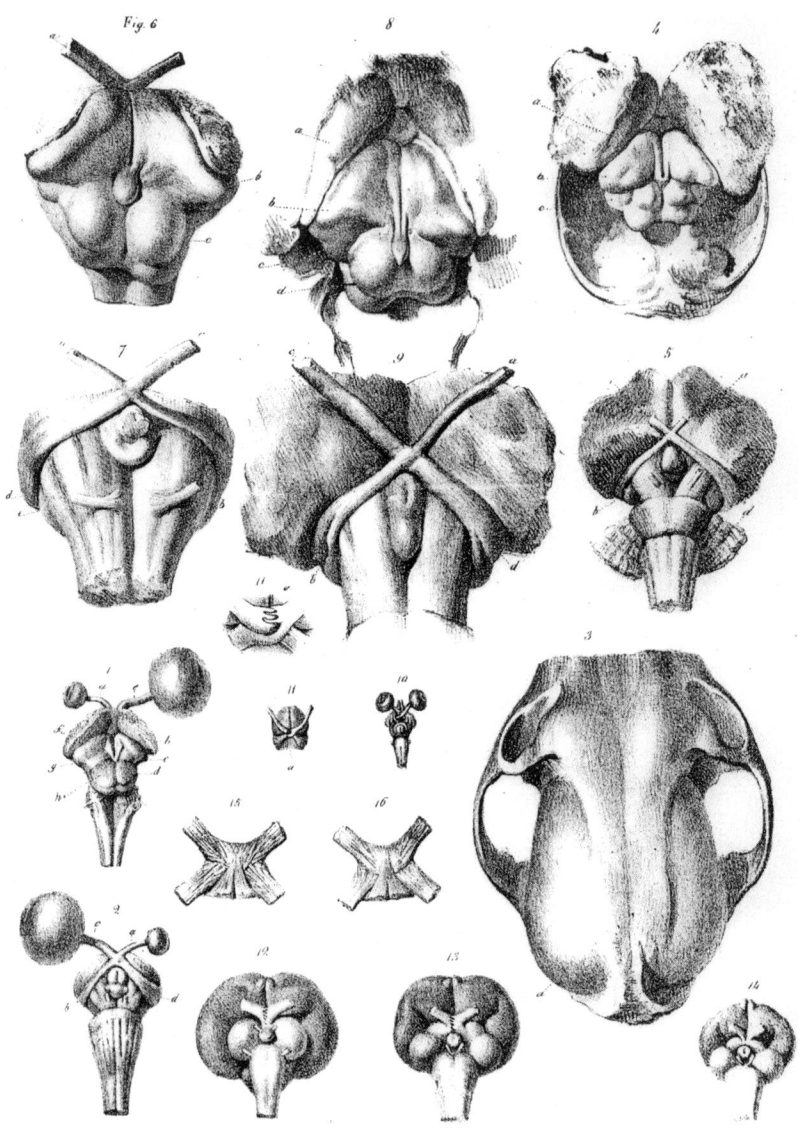

Figure 9.2
Panizza's Table IX. Figure 1. Brain of a rabbit that was blinded in the left eye when young. The corpus striatum, optic thalamus, and the left quadrigeminal eminence corresponding to the right eye are more developed (e–h). Vice versa for the opposite side (a–d). Figure 2. Inferior region of the same brain that shows the atrophy of one nerve relative to the other (a–d). Figure 3. Skull of a dog blinded in the left eye before it was weaned. After 1 year the skull showed an increase on the left side (a). Figure 4. Objects of the brain cavity of the above dog. We can see that the corpus striatum, optic thalamus, the geniculate body and the bigeminal eminence on the left are more pronounced, particularly the external contour of the posterior part of the optic thalamus: (a) corpus striatum, (b) optic thalamus, and (c) geniculate body. Figure 5. Inferior surface of the same brain, where you can see the difference in volume of the two nerves before and after the aja (a–d). Figure 6. Section of the brain of a horse that had lost the right eye many years earlier. The right optic nerve has atrophied, while the left is highly developed. In the objects in the ventricle the right optic thalamus was more developed and protruding in the rear (b), just as are the corresponding bigeminal eminence (c). Figure 7. Inferior surface of the same brain where you can see the great difference in volume between the two nerves (a–d), and how the geniculate body is pronounced (e). Figure 8. Section of a calf brain that was blinded in the left eye after an injury a few days after birth. Killed after 5 months, the left olfactory and optic thalamus were developed, as were the corresponding bigeminal eminence (a–d), while the opposite parts were atrophied. Figure 9. Differences between the two nerves (a–d) before and after the square aja of the above calf. Figure 10. Brain of the C. viridiflavus that serves to demonstrate that the two optic nerves at meeting in the aja form a fissure, where one passes in the other without mixing. Figure 11. Section of the chameleon brain that shows that each optic nerve when it meets the other divides into three fascia that within the fissure intertwine with the fascia of the other without mixing. They intertwine like the fingers of one hand with the other without mixing (a). Figure 12. Brain of a crow showing the decussation of the optic nerves. Figure 13. Brain of a duck showing the perfect decussation of the optic nerve. Figure 14. Decussation of the optic nerves in a chicken. Figure 15. Superior surface of the square aja of humans. Figure 16. Square aja shown from its inferior surface.

---

exist great diversity; between the inferior ones though the left was more developed.

Dissecting the brain of horses blinded in one eye, I was able to make various observations, one that enabled me to draw Figures 6 and 7 of Table IX. The brain was that of a horse blinded in the right eye: this optic nerve was very slender, ash-colored, gelatinous, whereas the other was large and very white. On the superior surface of the square aja you could see a small portion of the atrophied nerve running diagonally from the nerve coming from the other side and immersing itself like a wedge in its substance to exit from the opposite side. In the inferior surface of the aja at the join you could see a white streak that crosses diagonally, and similarly directed itself to the opposite side. On the superior surface of the aja we could see many fibers that originated from the tuber cinereum and extended directly from the posterior margin of the chiasm to the anterior margin where they partly inserted themselves into the atrophied nerve. Behind the chiasm one could see the ash-colored left optic nerve up to the thalamus, and its volume was greater than that of the nerve that was located to the right in front of the aja. There was also a difference in size between the two thalami; the right one was better developed, as was the inferior geniculate body and the right quadrigeminal eminence (Table IX, Figures 6a–c and 7c–e), as well as [the] lateral medullary column of the right peduncle of the brain and the external fibrous apparatus of the optic thalamus that extends to the posterior superior circumvolutions.

With respect to humans, I refer only to the case of the cadaver of a young girl of 18 years that at the age of 3 years, from a stone hitting her left eye, had lost sight in her eye because it had become atrophied. The skull associated with the right parietal lobe showed a clear depression along the occipital margin of the bone. The corresponding cerebral mass was also depressed. On dissection of the brain, it was found that the left corpus striatum was swelled along its whole body when compared with the right one, thus from the point of its major elevation it was distant two lines from the septum lucidum, the other one, on the other hand, that is, the left one, was only half a line away. This was even more clear: the left optic thalamus was also more developed; there was almost no difference in the quadrigeminal eminence. In the dissection of the objects of the ventricles we found greater consistency in the left ones. Of the optic nerves, the left one, from the eye to the chiasm, was ash-colored and atrophied; behind the aja, the right one was smaller than the left one, but was not of a different color.

To confirm how much the integrity of the optic thalamus and of the corpus striatum influences the organs of sight, I refer to two important facts that I uncovered. A human with a hot temperament, 60 years old, one day while eating had a fainting episode from which he recovered quickly and was able to categorically answer questions presented to him: the entire right part of the body was in a state of paralysis with respect to both senses and motion: the pupil of the right eye was dilated and immobile, and the sight was almost all lost. Placed in bed, and despite attention and care, the morbid symptoms continued to worsen, his intelligence continued to decrease, after 12 [hours] the sick human was completely apoplectic with complete paralysis of sense and motion on the right side of his body; he died on the fourth day. As soon as the problem began, having noticed blindness of the right eye, I predicted that the hemorrhage had to be located in the left optic thalamus. In the dissection we could find almost no infection of the meninges, and of the surface of the brain, little fluid in the ventricles, and instead a lot of blood was in the thickness of the left thalamus.

A friend of mine, that for a few days had been suffering from a headache, was struck after lunch with an apoplectic stroke, and although there was complete paralysis of the right side of his body, and perfect sight in the corresponding eye, his mind was free, but not his speech. He lived for 4 days maintaining his intellectual faculties, that only slightly began to vanish in the last 12 [hours] of life. In the sections we did not find accumulation of blood nor any fluid in any part of the brain, nor infection in the large or small blood supply. There appeared instead a softening of [the] left hemisphere and in particular in the optic thalamus. In the posterior and superior parts of the cerebral circumvolutions we found a very notable

softening, confirmed by all persons present at the examination of the cadaver, among which was my learned colleague and friend Professor Platner:
the comparison with the analogous parts in the other hemisphere left no
doubt in this regard. The quadrigeminal eminence and the optic nerve
were in a normal state.

To summarize what I have said so far on the origin of the optic nerve
in the three classes of animals, fish, birds, and mammals, the results showed
that:

1. In fish the formation of the optic nerve is contributed little by the
   anterior lobe, much by the hollow lobe, and within the internal
   objects of this the corpus striatum and the above placed eminence.

2. In the lobe and optic thalamus of birds, some fibrous fascia of the
   hemispheres alongside the peduncle join to the thalamus and,
   therefore, the optic nerve, as well as the radiating lamina of the internal wall of the cerebral hemisphere, lamina that with its nervous
   fascia intersect in part the optic nerve and in part the external extreme of the same optic thalamus (all contribute to the origin of
   the optic nerve).

3. In mammals the quadrigeminal eminence (especially the nates), the
   optic thalamus, the fibrous fascia that originate in the posterior
   cerebral circumvolutions, and also the tuber cinereum and the substance of the lateral wall of the infundibulum of the third ventricle
   (all contribute to the origin of the optic nerve).

From the things said up to now, what follows are some considerations
that should be noted.

If a bird, or a suckling just born, loses sight in one eye and lives then a
normal development, this will result in a major elevation of the skull and of
the brain on the side of the blind eye. If you look at Figure 20, Table VIII
[here, figure 9.1], that shows the head of a chicken in which I blinded the
left eye while it was still a chick of a few days old: killed after 1 year, the

skull showed an elevation on the left side, and I similarly found more prominent in the brain the hemisphere and the optic lobe on the left side, as is
seen in Figure 21, Table VIII, or better still in Figures 22a and b, of Table
VIII. The same experiment was performed on ducks a few days old that,
along with a perfect development, also showed a skull (Table VIII, Figure
24) that was more prominent on the left side, as well as a more prominent
left cerebral hemisphere (Table VIII, Figure 25a). Figure 26 (the inferior
surface of the same brain) shows considerable development of the right
nerve, and of the hemisphere and optic lobe on the left, while showing
atrophy of the left optic nerve and the right optic lobe. Following blindness
of the left eye on many lactating rabbits and dogs that were then sacrificed
after 1 year, I could reveal a more or less normal development of the skull
and brain corresponding to the good eye. The third figure of the ninth
Table shows the left part of the skull of a dog that is more developed (letter
a), as was the cerebral hemisphere maximally at the posterior lateral portion.
The internal parts can be seen in the fourth figure that shows the enlarge-
ment of the corpus striatum, of the left optic thalamus, of its external geniculate body, and of the bigeminal eminence (letters a–c). The fifth figure
shows the change of the optic nerve in front and also posteriorly to the aja.
I do not believe, however, that in birds one can replicate the major elevation of the cerebral hemisphere by just increasing the lobe or optic thalamus; a lot is due to the development of the fibrous apparatus of the optic
thalamus, that, uniting with the medullary fascia of the cerebral peduncle,
expands in the cerebral hemisphere, so much so that it is notable on the
side of the good eye as a major development of the optic thalamus and of
the corresponding corpus striatum; these two bodies should be considered
the center of the emanations of the fibrous apparatus that is part of the cerebral circumvolutions, particularly of the posterior part.

For the purposes of a demonstration, let us make, in the brain of a
human, dog, or rabbit, a longitudinal cut that divides the corpus callosum
into two symmetrical parts and continue to cut to the base of the brain so
as to separate the brain into two equal portions. In each of these we lift up

the corresponding part of the corpus callosum and vault with three columns so as to uncover the external contours of the corpus striatum, the optic thalamus, and of the digital cavity; then we graze with a knife blade the medullary substance outside the margins indicated above and immediately one will see appear the considerable nervous fascia exiting from the aforementioned bodies; fascia that by spreading they extend toward the periphery of the cerebral hemispheres, then to their circumvolutions, and those of the posterior contours of the optic thalamus to the posterior superior circumvolutions. It is the existence of these nervous fascia that suggests an explanation for the above phenomenon, that even a light lesion to the periphery of the cerebral hemisphere, if it reaches the fibrous substance, always results in blindness of the opposite eye, or complete blindness if the same lesion is made in both hemispheres, without causing disorders of other cerebral functions.

The above observations suggest another consideration, that is, that one part of the brain located even at the base can give indication of its greater swelling in the superior surface of the skull if this swelling occurs while the brain is still developing. In [the case of] swelling, for example, of the optic lobe, the thalamus can do no less than occupy the space that belongs to the overlying part, and this should be pushed above the normal limits so that, coming in contact against the flexible walls of the skull, [it] will join hand-in-hand to alter its good form, thus breaking its symmetrical disposition. Therefore, if it is right to consider the brain as a complex of organs, it is not as easy to determine their number, the seat of each one, and their respective functions, and I believe that in this serious subject we are a long way from having an absolute and decisive language. In my opinion, in-depth studies are still required on the brain organization of humans and brutes; and these should be carried out considering the development and improvement of its parts, and the relation of these to the various intellectual and moral manifestations. In fact, we should continue emphasis on anatomical studies, and especially on the comparative examination of the brain; carrying out careful studies on live animals with interest in the various brain parts, observing closely what happens to them, and considering the various pathological cases, we may be able to clarify some of the many principles of the physiology of the brain that are still unclear, and we can better reach this goal, when we make more scrupulous and serious observations of all principles investigated.

From the research carried out on the origins of the optic nerve, of that which is described in this paper, the results demonstrated that the optic nerve, in contrast to other cranial nerves, rather than having a unique origin, is composed of fibers originating from various points of the brain mass, of which the more important ones find themselves far from the openings at which the nerve is destined to exit from the skull, thus it must extend over a considerable section of this cavity, along which it comes into contact with various objects forming the cerebral mass. From this we could argue that, regarding a nerve that has an important part in intellectual development, that furnishes many precious materials in this exercise above all other noble faculties, to be, as it has been said, the mirror of the soul, nature wanted to put it in contact with a great number of the organs required for intelligence. We could object that another nerve, that has a duty no less important, for example the acoustic nerve, originates from a singularly narrow tract of the brain, while everybody knows that in this point regarding the origin of the acoustic nerve the fibers come from more noble parts of nervous centers, such that this nerve collects information from other organs, which are tied to the life and all of its attributes. But, also disregarding these arguments, one cannot deny that the various origins of the optic nerve explains the observed phenomenon in birds as well as mammals, that after an injury of even a small and superficial part of the brain at the anterior extremity and especially posteriorly at the corpus striatum or the optic thalamus, the sight of the opposite eye is always damaged, without any other damage to motion, the senses, and intelligence.

The research on the optic nerve gave me new motivation to examine closely the manner in which the optic nerve behaves in the chiasm, in other words how the two nerves unite. We know that some opinions are that the

---

optic nerves do nothing but connect one to the other; others admit total crossing, that is, the one on the right side crosses to the left side after the chiasm. Many others, and they are the most, maintain that there is only a partial decussation, a crossing of the major parts of the fibers, and especially of the internal ones, that pass to the opposite sides, while the external fibers of each nerve continue their path without penetrating into the aja, favoring only the curve of the external side of the nerve.

Generally, in fish the optic nerves overlap one another, remaining united with the thick cellular texture. In some reptiles, such as *Coluber virid-flavus*[25] (Table IX, Figure 10), each nerve divides into two fascia, one that passes in the middle of the hole left by the other. In the brain of the chameleon (Table IX, Figure 11) at the crossing the nerve divides into small fascia, which truly entwine with the small fascia of the other nerve without mixing, like what happens when you cross the fingers of the hand (Table IX, Figure 11a). In birds, such as the crow, chickens, and in palmipeds, clearly you can see the decussation and the interlacing of the right fibers with the left ones (Table IX, Figures 12–14). These observations are confirmed by the examination of the pathological pieces collected from chickens in which one eye was blinded.

While the total decussation is clear in the above-mentioned animals, examining this point is difficult in mammals and, therefore, in humans. In the works of many learned authors we find the representative figure, in which [we] see the partial crossing of the internal fibers of each nerve, while the external ones follow the primitive way directing themselves to the corresponding orbit without mixing with the fibers of the other. After a close examination of a series of preparations, I am able to accept that the external margin of the aja does not consist of portions of the nerve that continues, but from fibrous fascia that are directed, some superficial, others deep, but in a way that does not continue to the external margin of the nerve. In fact I noticed a completely different direction, and, therefore, I traced Figures 15 and 16, Table IX. As a result of this, at the joining of the optic nerve at the aja, some of its fascia, particularly the deep ones, cross diagonally and emerge from the opposite side, the others run along the external margin, but sooner or later they wrap around it, to also end up on the opposite side, therefore, completing a perfect decussation: in other words I saw that from the two lateral lamina of the infundibulum of the third ventricle exit some medullary fascia that expand into the superficial surface of the aja directing themselves from back to front, and in this way at the inferior surface of the aja I noticed longitudinal fibers that derive from the tuber cinereum. These anatomical observations favor perhaps the opinion that there is a complete decussation of the optic nerve even in humans, opinions that make it easy to understand the pathological phenomenon noted by all, that after cerebral damage you almost always have perfect blindness of one eye, while the other remains perfectly normal, a phenomenon that would remain inexplicable given just a partial decussation.

On the richness and distribution of the blood vessels in the chiasm, and the particularities of the structure of the optic nerves, up to the square aja and after, with respect to the form and the distribution of the large and small nervous granulations, I will speak on another occasion.

Notes

This paper was originally published in *Behavioural Brain* Research (58: 529–539 [2002], M. Colombo, A. Colombo, and C. G. Gross, "Bartolomeo Panizza's *Observations on the Optic Nerve [1855]*"). Dr. M. Colombo is in the Department of Psychology and the Centre for Neuroscience, University of Otago, Dunedin, New Zealand, and A. Colombo is his father. Professor M. Bentivogio, University of Verona, helped with the older Italian terms.

1. Broca, 1960; Fritsch and Hitzig, 1960.

2. Panizza, 1855. This was republished as Panizza, 1856, and it was from the latter that the two sets of illustrations (labeled in the original and referred to in the text as Table VIII and Table IX, here shown as figures 9.1 and 9.2) were reproduced.

3. Finger, 1994; Gross, 1998a.

---

4. Gross, 1998a, 1999d.

5. Gross, 1998a, 1999d; Young, 1970.

6. Broca, 1960.

7. Fritsch and Hitzig, 1960

8. Gross, 1998a; Young, 1970.

9. Manni and Petrosini, 1994.

10. Manni and Petrosini, 1994; Mazzarello and Della Sala, 1993; Zago et al., 2000.

11. Von Gudden, 1870.

12. Munk, 1878; Luciani and Tamburini, 1879; Munk, 1881.

13. Tamburini, 1880; Polyak, 1957.

14. Finger, 1994; Polyak, 1957; Manni and Petrosini, 1994; Mazzarello and Della Sala, 1993; Zago et al., 2000.

15. Zago et al., 2000.

16. Panizza, 1855.

17. Throughout the manuscript Panizza refers to "eminenze bigemine." Ranson, 1920, states: "In the lower vertebrates there are but two elevations in the roof, the optic lobes or corpora bigemina, and these, which correspond in a general way to the superior colliculi, are visual centers" (p. 165). It is clear that the "eminenze bigemine" refer to the optic tectum in fish, also known as the superior colliculi in mammals. To maintain as close a wording as possible to the original document we have translated "eminenze bigemine" into "bigeminal eminence."

18. In most nonmammalian vertebrates, the third ventricle extends into the optic lobe, where it is referred to as the tectal ventricle (see Butler and Hodos, 1996). Hence, "hollow lobe" refers to the optic tectum.

19. Medullary refers to white matter.

20. The term "corpo striato" refers to the "corpus striatum," which consists of the caudate nucleus, putamen, and globus pallidus. The fact that at one time the optic thalamus

—————

and the corpus striatum were considered both part of the basal ganglia (Whitaker, 1887; Whitehead, 1900), and the fact that they lie in close proximity to each other in the brain, might explain why in many cases Panizza makes reference to either the optic thalamus or corpus striatum.

21. The term "eminenze quadrigemelle" refers to the superior and inferior colliculi. The actual translation would be "corpora quadrigemina." Again, to maintain as close a wording as possible to the original document, we have translated "eminenze quadrigemelle" to "quadrigeminal eminence."

22. Aja and aja quadrata refer to the immediate area around the optic chiasm. We have translated "aja quadrata" into "square aja."

23. Nates are buttocks and refer to the superior colliculi. It was common to name brain parts after sexual organs they supposedly looked like. Whitaker (1887) states, "The corpora Quadrigemina are four rounded tubercles separated from each other by two grooves, the one longitudinal, the other transverse. They are placed in pairs, on each side of the middle line, behind the pineal gland and above the aqueduct of Sylvius; the anterior pair are called the nates, the posterior pair, the testes." The testes in this case are the inferior colliculi.

24. A line is approximately 1/12 of an inch.

25. A member of the colubridae family also known as a common racer snake.

10

JOSEPH ALTMAN AND ADULT NEUROGENESIS: THE DOGMA OF "NO NEW NEURONS" IN THE ADULT MAMMALIAN BRAIN

From the beginning of the neuron doctrine in the late nineteenth century until the early 1990s a central dogma in neuroscience was that "no new neurons are added to the adult mammalian brain."[1] By the end of the nineteenth century, the idea that the brain of the adult mammal remains structurally constant was already universally held by the neuroanatomists of the time. Koelliker, His, and others had described in detail the development of the central nervous system of humans and other mammals.[2] They found that the structure of the brain remained fixed from soon after birth. Because the elaborate architecture of the brain remained constant in appearance, the idea that neurons were continually added to it was, understandably, inconceivable. Similarly, Ramón y Cajal and others had described the different phases in the development of the neuron, terminating with the multipolar structure characteristic of the adult.[3] As neither mitotic figures nor preadult developmental stages had been seen in the adult brain, the possibility of continuing neuronal addition to the adult brain was rarely, if ever, seriously entertained. As Ramón y Cajal put it, "In the adult centers the nerve paths are something fixed, ended and immutable. Everything may die, nothing may be regenerated."[4]

In the first half of the twentieth century, there were occasional reports of postnatal neurogenesis in mammals but these were usually ignored by textbooks and rarely cited.[5] Presumably this was because of the weight of authority opposed to the idea and the inadequacy of the available methods both for detecting cell division and for determining whether the apparently new cells were glia or neurons.[6]

ALTMAN CHALLENGES THE DOGMA AND IS IGNORED

An important advance in the study of neurogenesis came in the late 1950s with the introduction of [^{3}H]-thymidine autoradiography. [^{3}H]-thymidine is incorporated into the DNA of dividing cells. Therefore, the progeny of cells that had just divided could be labeled, and their time and place of birth determined. Initially, this new method was applied exclusively to the study of developing rodents.[7] The emphasis on using this method to study pre- and perinatal development, rather than looking across the life span of the animal, reflected the persistence of the belief that neurogenesis did not occur in the adult mammal.

Starting in the early 1960s, Joseph Altman (b. 1925) challenged this idea of "no new neurons in the adult brain." He published a series of papers reporting thymidine autoradiographic evidence for new neurons in the dentate gyrus of the hippocampus, the olfactory bulb, and the cerebral cortex of the adult rat.[8] He also reported new neurons in the neocortex and elsewhere in the adult cat.[9] (See figure 10.1.) Most of the new neurons were small, and Altman suggested that they were crucial for learning and memory.[10] Although published in the most prestigious journals of the time, such as the *Journal of Comparative Neurology*, *Science*, and *Nature*, these findings were totally ignored or dismissed as unimportant for over two decades. As late as 1970, an authoritative textbook of developmental neuroscience stated that "there is no convincing evidence of neuron production in the brains of adult mammals."[11]

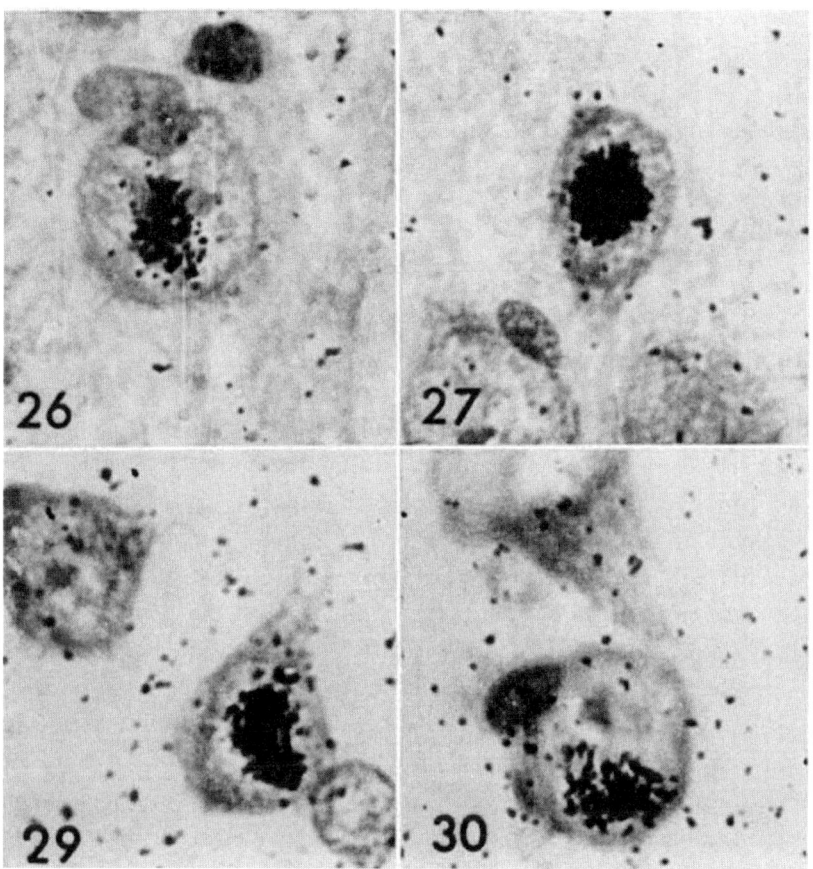

Figure 10.1
Autoradiograms of apparently labeled neocortical (lateral gyrus) neuron nuclei in an adult cat that was injected intraventricularly with [³H]-thymidine cresyl violet stain (Altman, 1963).

Altman was not granted tenure at MIT and moved to Purdue University where he eventually turned to more conventional developmental questions, perhaps because of the lack of recognition of his work on adult neurogenesis. Unable to get grants, he supported his work by producing magnificent brain atlases.[12]

KAPLAN CONFIRMS ALTMAN AND IS ALSO DISMISSED

Fifteen years after Altman's first report, direct support for his claim of adult neurogenesis came from a series of electron microscopy studies by Michael Kaplan and his coauthors. First, they showed that [³H]-thymidine-labeled cells in the dentate gyrus and olfactory bulb of adult rats have the ultrastructural characteristics of neurons, such as dendrites and synapses, but not of glia (astrocytes or oligodendrocytes).[13] Then Kaplan reported autoradiographic and ultrastructural evidence for new neurons in the cerebral cortex of adult rats, again confirming the earlier claims of Altman.[14] Finally, he showed mitosis in the subventricular zone of adult macaque monkeys by again combining [³H]-thymidine labeling and electron microscopy.[15] During this period, Kaplan was, successively, a graduate student at Boston University, a postdoctoral fellow at Florida State University, and an assistant professor at the University of New Mexico. Attacked for his iconoclastic claims, Kaplan left the field, became a medical student, and now works in rehabilitation medicine.[16] In spite of his evidence for adult neurogenesis, Kaplan's work had little effect at the time, as measured by citations or follow-up studies. Again, as in Altman's case, publication in prestigious and rigorously reviewed journals, such as *Science*, the *Journal of Comparative Neurology*, and the *Journal of Neuroscience* by an unknown figure was not sufficient to make any marked dent in the dogma.

An important reason for the small impact of Kaplan's work may have been a study presented at a meeting in 1984 and published the following year. Pasko Rakic, the author of the study, was (and still is) professor at Yale Medical School and arguably the leading student of primate brain development. He carried out a [³H]-thymidine study of adult rhesus monkeys in which he examined "all major structures and subdivisions of the brain including the visual, motor, and association neocortex, hippocampus, [and] olfactory bulb." Rakic found "not a single heavily labelled cell with the morphological characteristics of a neuron in any brain of any adult animal" and concluded that "all neurons of the rhesus monkey brain are generated during prenatal and early postnatal life."[17]

Rakic's papers had a profound influence on the development of the field.[18] Subsequent work in adult rhesus monkeys by Eckenhoff and Rakic, using a combination of thymidine autoradiography and electron microscopy, also failed to find new neurons in the adult.[19] Furthermore, the authors questioned the reports of adult neurogenesis in rats with the suggestion that rats never stop growing and so never become adults. (In fact, there are strains of rats that do stop growing and also show adult neurogenesis,[20] but this was not known at the time.) For Eckenhoff and Rakic, the supposed lack of adult neurogenesis in primates made sense, because "a stable population of neurons may be a biological necessity in an organism whose survival relies on learned behavior acquired over a long period of time."[21] Furthermore, Rakic suggested that the "social and cognitive behavior" of primates may require the absence of adult neurogenesis.[22]

Humans often show a basic need to distinguish themselves from other animals, and their order—primates—from other orders on cognitive grounds.[23] Although neuroscientists have often tried to make these distinctions in terms of brain structure or function, Rakic's suggestion may be the only time that the social and cognitive differences between primates and nonprimates was attributed to the presence or absence of adult neurogenesis and, more generally, to the structural stability of the brain.

There were three developments that led to a vindication of Altman's pioneering work and to general acceptance that new neurons are added to the adult mammalian brain and that this was probably an interesting and important phenomenon. The first development was the demonstration of neurogenesis in adult birds. The second was the introduction of new methods for labeling new cells and for distinguishing neurons from glia. Finally, demonstrations that neurogenesis could be up- and downregulated by important psychological variables such as stress, environmental complexity, and learning raised the possibility that adult hippocampal neurogenesis might be important for cognition in higher animals.

AVIAN NEUROGENESIS

Starting in the late 1980s, Nottebohm and his colleagues at Rockefeller University began a systematic analysis of the neural basis of song learning in birds. They discovered a set of brain mechanisms that are crucial for bird song and showed how the volume of two nuclei were a function of variables such as sex, sexual maturity, song complexity, species, testosterone level, and season.[24] The seasonal and hormonal changes in the volume of these song-related nuclei were so great in some species that Nottebohm set out to examine the possibility that these changes were due to fluctuations in the actual number of neurons in the adult avian brain.

In a series of elegant experiments, Nottebohm and his colleagues showed that, indeed, thousands of new neurons are added every day to the avian brain. They did so by, first, showing the production of new cells with thymidine labeling;[25] second, producing ultrastructural evidence that the new cells were neurons receiving synapses;[26] and last, in a technical tour de force, showing that the putative neurons responded to sound with action potentials.[27] In subsequent studies, they showed that the axons of new neurons extended over long distances, that neuronal birth and death proceeded in parallel, that in both singing and nonsinging species neurogenesis was widespread throughout the avian forebrain—including the hippocampus—and that in the latter structure it was modulated by environmental complexity and learning experience.[28]

In spite of this unassailable evidence of neurogenesis in parts of the adult bird brain known to be homologous to primate cerebral cortex and primate hippocampus, these studies tended to be viewed as irrelevant to the primate or even the mammalian brain. Rather, the evidence for avian neurogenesis was viewed as an exotic specialization related to the necessity for flying creatures to have light cerebrums and to their seasonal cycles of singing.

NEW TECHNIQUES FOR DETECTING NEUROGENESIS

Beginning around the 1990s, there were several developments that finally established the reality of neurogenesis in the dentate gyrus of the adult rat. One was the demonstration that the new cells in the rat dentate gyrus extend axons into the mossy fiber pathway.[29]

The second important development was the introduction of the synthetic thymidine analog BrdU (5-bromo-3'-deoxyuridine). Like thymidine, BrdU is taken up by cells during the S-phase of mitosis and is a marker of proliferating cells and their progeny. BrdU labeling can be visualized with immunocytochemical techniques and does not require autoradiography.[30] More recently, an endogenous marker for cell proliferation, Ki-67, was introduced. Ki-67 is a protein that is a cellular marker for cell proliferation. It is present during mitosis but is absent in the resting cell.[31]

Perhaps the most important advance was the use of cell-type-specific markers enabling the immunohistochemical distinction of the newly generated neurons from glia cells. Among the markers for mature neurons are NSE, MAP-2, TuJ1, and NeuN. Although some of these markers have been shown to stain nonneuronal cells under certain conditions and others do not stain all neuronal types,[32] the expression of several of these antigens in a population of adult-generated cells is considered good evidence that new neurons have been produced. There are also markers for immature neurons and for glia (oligodendrocytes and astrocytes). The combination of BrdU for detecting new cells with immunochemical markers for neurons now allowed the identification of new neurons. Other markers for new neurons are now available.[33]

REGULATION OF NEUROGENESIS

The advent of these new techniques meant that by the 1990s, Altman's claim that new neurons were added to the adult dentate gyrus had been confirmed several times, and by now it is well established for a variety of mammals including humans and other primates.[34]

But was this phenomenon more than some ontogenetic lag or phylogenetic vestige? At least in rats, the finding that the number of new hippocampal cells is so large, over 9,000 cells per day, most of which are neurons, makes this very unlikely.[35] Furthermore, dentate gyrus neurogenesis in the rodent can be modulated by a number of experiential variables and so might be important for cognitive function.[36] For example, acute and chronic stress decreases hippocampal neurogenesis. Adrenal steroids probably underlie this effect as stress increases adrenal steroid levels and glucocorticoids decrease the rate of neurogenesis. By contrast, there are several conditions that increase the number of adult-generated dentate gyrus cells, environmental complexity and wheel-running being particularly well studied enhancers of adult neurogenesis.

In Altman's earliest studies he speculated that adult neurogenesis might play a crucial role in learning and memory.[37] In recent years this idea has been subjected to an increasing amount of experimental examination. Although there are conflicting results, the preponderance of evidence supports Altman's speculation: the number of new neurons often positively correlates with learning of hippocampal-dependent tasks: learning such tasks tends to increase the number of new neurons and the depletion of new hippocampal neurons is reported to impair hippocampal-dependent learning.[38]

ADULT NEUROGENESIS IN THE OLFACTORY BULB AND CEREBRAL CORTEX

At about the time that Altman's finding of neurogenesis in the dentate gyrus was confirmed, his report of neurogenesis in the adult olfactory bulb was also replicated.[39] Neurogenesis in the adult olfactory bulb has now been shown for a variety of mammals, including humans,[40] and, at least in rats, it is modulated by olfactory experience and learning.[41]

The status of Altman's report of adult neurogenesis in the cerebral cortex is less clear. Beyond Altman and Kaplan's work, a number of investigators have reported cortical neurogenesis in the hamster, rat, marmoset, and macaque cortex. However, others have failed to find cortical neurogenesis. Gould and Cameron and Dayer have reviewed the positive and negative studies and suggested that the negative results were due to insufficient sensitivity of the methods used and the small number and size of the new cortical neurons.[42] (See this chapter's postscript).

Why Were Altman's Discoveries Ignored for Almost 30 Years?

There appear to be several reasons why Altman's discovery of neurogenesis in the hippocampus and the olfactory bulb was ignored. First, there were not accessible and reliable techniques for the objective differentiation of small neurons from glia, particularly astrocytes. Until the 1990s this distinction could be made only by "an expert eye," and almost by definition, "experts knew" that adult neurogenesis did not occur in mammals. Another reason was that Altman, although in a leading university (MIT), was at the time of his early adult neurogenesis papers a junior faculty member in a psychology department and had not been trained in a distinguished, or indeed any, developmental laboratory or one using autoradiographic techniques. Finally, the dogma of "no new neurons" was universally held and vigorously defended by the most powerful and leading primate developmental anatomists of his time.

The continued resistance to acceptance of neurogenesis in the adult cerebral cortex may be due, in part, to the much lower incidence of cortical neurogenesis than hippocampal neurogenesis and therefore the greater importance of sensitive methods for detecting new neurons there. It may also reflect the continued investment of more traditional members of the community in denying neuronal plasticity.

POSTSCRIPT

COMPARISON OF FOUR WHO WERE "BEFORE THEIR TIME"

In my previous *Tales in the History of Neuroscience* I described a more extreme case of someone "before his time"—Emmanuel Swedenborg (1688–1772), who anticipated sensory and motor function of the cortex and, arguably, even the neuron theory by over 100 years.[43] Yet his writings on brain function remained unknown until the twentieth century, by which time many of his ideas had been confirmed. There are both common elements and ones that were very different in the neglect by their contemporaries of Swedenborg's ideas on the cortex, Bernard's dictum on the constancy of the internal environment (chapter 8), Panizza's discovery of visual cortex (chapter 9), and Altman's discovery of neurogenesis.

Swedenborg, Panizza, and Altman faced impregnable ideological resistance: for Swedenborg, the dogma that the cortex was a functionless rind, for Panizza the dogma that the cortex had "higher functions" but neither sensory nor motor ones, and for Altman the dogma of no new neurons.

Swedenborg's publications never reached the scientific community until they were long outdated. Panizza's paper was published in a local journal but one that circulated to major scientific societies (exactly as was the case for Gregor Mendel). By contrast both Bernard's and Altman's were very widely available.

Bernard was the most famous French scientist of his time (and arguably, all time); Panizza was very highly regarded in his university, Altman was at a distinguished research institution whereas Swedenborg was not even recognized as a biologist.

All four had ideas that were difficult or impossible to test in their own time.

It took over 150 years for Swedenborg to be rediscovered. Bernard's ideas on the internal milieu took about 50 years to be understood. Panizza was cited only after his discovery was rediscovered, and then was returned to his previous totally obscure status. Altman's adult neurogenesis findings were ignored for about 30 years and are now ubiquitous in the opening paragraphs of papers on adult neurogenesis.

They were all iconoclasts and their icons were resilient.

Adult Neurogenesis in the Cerebral Cortex Today

By 2008 Altman's reports of adult neurogenesis in the dentate gyrus of the hippocampus and in the olfactory bulb had been repeatedly confirmed with a variety of techniques and in many mammals including humans. By contrast his claim for adult neurogenesis in the cerebral cortex has remained a subject for debate and controversy. Since 1999 there have been at least five groups of investigators finding positive evidence for adult cortical neurogenesis in rodents and primates and about the same number failing to do so.[44]

There are several possible reasons for the difficulty in replicating the positive findings of cortical neurogenesis. First, the new cortical neurons that have been reported are much harder to detect than those in the olfactory bulb and hippocampus because they are much rarer and distributed unevenly over a much larger volume of tissue. Second, the new cortical neurons are interneurons much smaller than the typical cortical pyramidal cells among which they are sparsely scattered.[45] Third, in many of the negative studies, histological protocols were used that have been shown to tend to reduce tissue quality and diminish the signal-to-noise ratio for detecting new neurons. Finally, and perhaps most important, the studies that failed to find cortical neurogenesis used techniques that also failed to show adequate evidence for new neurons in the dentate gyrus, presumably because of their poor sensitivity or tissue destructiveness. For inferences to be made about failure to find cortical neurogenesis it is necessary to have the positive control of good evidence for hippocampal neurogenesis with the same techniques, and this has not yet been the case.[46] New, more sensitive techniques and the use of positive controls are needed to resolve this issue.

It should be noted that the very small reported incidence of new cortical neurons, namely 1–2 cells per mm^3, should not belie the possible importance of these cells. At least in sensory systems, the activity of a single or small number of neurons has been shown to influence behavior.[47]

In recent years adult neurogenesis has also been reported in the striatum, amygdala, hypothalamus, substantia nigra, and brain stem in rodents or primates or both.[48] Some of these observations have been confirmed but others have not, perhaps because of the sensitivity issues discussed for cortex.

Altman's speculation that the new neurons in the hippocampus might play a role in learning and memory has belatedly resulted in a burgeoning literature on the subject.[49] Since my paper was published there has been considerable additional evidence of the involvement of new hippocampal neurons in learning and memory tasks that require the hippocampus, such as spatial learning and eye-blink conditioning.[50] However, the task and timing parameters for the involvement of new hippocampal neurons are not well understood and seeming contradictions in the literature abound. The role of the new neurons in the processes underlying learning and memory remains totally obscure.

ALTMAN'S REFLECTIONS

As this book was going to press at the end of 2008 I received a lengthy "in press" typescript from Joseph Altman entitled "The discovery of adult mammalian neurogenesis."[51] Altman wrote:

...my recollection is that [my early publications] were not ignored at all but created considerable publicity...with the generous financial support that we were receiving...we were able to pursue our research goals and disseminate the data. ...But then things started to change in the late 1960s, although it took me several more years to realize that something was amiss. The first wake-up call came when I was supposed to be granted tenure at MIT and my promotion was denied.... I now know that there was a concerted attempt by some influential members of the neuroscientific community to marginalize us, but at the time I did not pay much attention to it.... This neglect of our work continued during the 1980s....

I am not aware of any public criticism or rebuttal of the data we have presented.... Instead of open criticism, there appears to have been a clandestine effort by a group of influential neuroscientists to suppress the evidence we have presented and, later on, to silence us altogether by closing down our laboratory. I make this accusation for the following reasons: (a) by the early 1980s we were starting to have difficulties in getting our grant applications approved; (b) by the mid-1980s we lost *all* our grant support; and (c) by the early 1990s we had several of our submitted papers outright rejected.... In fact, we have never since been able to get our grant applications approved, not even when the topic of our research shifted from adult neurogenesis to the comprehensive embryological study of CNS development in rats and, later, in humans.[52]

Under these conditions how did Altman's "laboratory survive, indeed prosper, without research grants and...remain productive while facing peer indifference and outright hostility?" One explanation Altman gives was that he had accumulated a "large collection of processed brains...[and could] analyze...[them]...and, instead of submitting papers to mainstream journals (a costly and painful enterprise)...started to summarize...results in books that also yielded some royalty." He continues:

> But there was also a psychological factor: namely, that I was well prepared by my earlier life experiences not to be discouraged by public indifference and hostility. When I committed myself to the study of brain-behavior relationships as a teenager...I was barred from finishing my formal education in quasi-fascist Hungary because I belonged to a disenfranchised religious/ethnic minority. When the Wehrmacht occupied Hungary, I was incarcerated in a forced-labor camp and worked on a railroad gang. After some time there, I escaped and lived clandestinely in Budapest, which very soon came under siege. My survival was aided, I tend to believe, by a syndrome that I developed, what I call "aparanoia." Blissfully neglecting the fact that they were after my kind, I did not hide but walked with head erect through the streets, past bayoneted soldiers and gendarmes looking for Jews and deserters. I convinced myself that I was going to survive no matter what the Nazis' intentions were. When the Red Army liberated us in 1945, I resumed my formal education. However, I could not tolerate the rising Communist dictatorship and fled the country in 1946. I became a stateless, displaced person in West Germany, waiting there for many years before I could secure the requisite documents to migrate to Australia. ...I was able to attend lectures and seminars at a prominent German university and spent endless hours in its library. I continued informally with my education in Australia, where I worked through the first half of the 1950s as a college librarian.

Altman then came to the United States in 1955 and, after pre- and postdoctoral training in New York City, went to MIT.

———

Notes

Some of the material in this chapter was published previously, in *Nature Reviews Neuroscience* (1: 67–73 [2000], "Neurogenesis in the adult brain: Death of a dogma") and in *Experimental Brain Research* (192: 321–334 [2009], "Three before their time"). The postscript has been added for this volume.

1. Ramón y Cajal, 1928; Rakic, 1985a; Jacobson, 1970.

2. Koelliker, 1896; His, 1904; Ramón y Cajal, 1999; Gross, 2000a.

3. Ramón y Cajal, 1928, 1999.

4. Ramón y Cajal, 1928.

5. Gross, 2000a.

6. Ramón y Cajal, 1928.

7. Sidman et al., 1959.

8. Altman, 1962, 1963, 1967, 1969; Altman and Das, 1965, 1966a, 1966b.

9. Altman, 1963.

10. Altman, 1967.

11. Jacobson, 1970.

12. E.g., Altman and Bayer, 1995, 1996; Bayer and Altman, 2007.

13. Kaplan and Hinds, 1977; Kaplan, 1984.

14. Kaplan, 1981; Altman, 1963; Altman and Das, 1966b.

15. Kaplan, 1983, 1985.

16. Kaplan, 2001.

17. Rakic, 1985a, 1985b.

18. Rakic, 1985a, 1985b.

19. Eckenhoff and Rakic, 1988.

20. Boss et al., 1985; Kuhn et al., 1996.

21. Eckenhoff and Rakic, 1988.

22. Rakic, 1985a.

23. Gross, 1993a, 1993b.

24. Nottebohm, 1985, 1989.

25. Goldman and Nottebohm, 1983.

26. Burd and Nottebohm, 1985.

27. Paton and Nottebohm, 1984.

28. Nottebohm, 1996; Barnea and Nottebohm, 1994, 1996; Kirn and Nottebohm, 1993.

29. Stanfield and Trice, 1988.

30. Nowakowski et al., 1989.

31. Scholzen and Gerdes, 2000.

32. Mullen et al., 1992; Deloulme et al., 1996; Sensenbrenner, Lucas, and Deloulme, 1997.

33. Perera, Park, and Nemirovskaya, 2008.

34. Perera, Park, and Nemirovskaya, 2008; Cameron et al., 1993; Seki and Arai, 1999; Abrous, Koehl, and Le Moal, 2005; Eriksson et al., 1998; Gould et al., 1998, 1999a.

35. Cameron and Mackay, 2001.

36. Gould, 2006.

37. Altman, 1967.

38. Abrous, Koehl, and Le Moal, 2005; Leuner, Gould, and Shors, 2006.

39. Lois and Alvarez-Buylla, 1994; Corotto et al., 1994.

40. Curtis et al., 2007.

41. So et al., 2008; Mandairon et al., 2006.

———

42. Gould, 2007; Cameron and Dayer, 2008.

43. Gross, 1998a, chapter 3.

44. The positive ones include Gould et al., 1999b, 2001; Dayer et al., 2005; Bernier et al., 2002; Runyan et al., 2006; and the negative ones include Kornack and Rakic, 1999, and Bhardwaj et al., 2006. These and other positive and negative results are examined in Gould, 2007; and Cameron and Dayer, 2008.

45. Gould, 2007; Cameron and Dayer, 2008.

46. Other technical reasons why some studies have been unable to find new cortical neurons are discussed in Gould, 2007. For a discussion of the effect of different methods in assessing cortical neurogenesis, see www.princeton.edu/~goulde/protocols (accessed September 17, 2008).

47. E.g., Brecht et al., 2004; Shadlen et al., 1996.

48. Gould, 2007.

49. Altman, 1967.

50. E.g., Bruel-Jungerman et al., 2007; Anderson et al., 2007; and Shors, 2008.

51. This has now been published (Altman, 2009)

52. E.g., Altman and Bayer, 1995, 1996; Bayer and Altman, 2007.

11

Donald R. Griffin: Echolocation and Animal Consciousness

Most scientists seek—but never attain—two goals. The first is to discover something so new as to have been previously inconceivable. The second is to radically change the way the natural world is viewed. Don Griffin did both. He discovered (with Robert Galambos) a new and unique sensory world, echolocation, in which bats can perceive their surroundings by listening to echoes of ultrasonic sounds that they produce. In addition he brought the study of animal consciousness back from the limbo of forbidden topics to make it a central subject in the contemporary study of brain and behavior.

Early Years

Donald Redfield Griffin (1915–2003) was born in Southampton, New York, but spent his early childhood in an eighteenth-century farmhouse in a rural area near Scarsdale, New York. His father, Henry Farrand Griffin, was a serious amateur historian and novelist who worked as a reporter and in advertising before retiring early to pursue his literary interests. His mother, Mary Whitney Redfield, read to him so much that his father feared for his ability to learn to read. His favorite books were Ernest Thompson

Seton's animal stories and the *National Geographic's Mammals of North America*. An important scientific influence on the young Griffin was his uncle Alfred C. Redfield, a Harvard professor of biology who was also a bird-watcher, hunter, and one of the founders of the Woods Hole Oceanographic Institution.

Young Griffin's boyhood hobbies were to become the central core of his professional interests and achievements. By the age of 12 Griffin was trapping and skinning small local mammals. Because of his poor teeth, his parents regularly took him to a Boston dentist. These trips were rewarded with visits to the Boston Museum of Natural History, where its librarian introduced him to scientific journals, and its curators to turning his trapped animals into study skins. At 15, with his uncle's encouragement, he subscribed to the *Journal of Mammalogy* where he was to publish five papers before graduating from college. In his autobiographical writings, Griffin described his schooling as "extraordinarily irregular." After he spent a few years at local private schools his "long-suffering" parents decided on home schooling. His father taught him English, history, Latin, and French. A former high school teacher handled the German and math. After a few years of trapping, skinning, sailing, and a couple of hours of daily lessons, his parents sent him to Phillips Andover, where he started but never finished the tenth and eleventh grades. The next year was spent at home again, collecting and sailing, and with tutoring adequate that he was admitted to Harvard College in the fall of 1934.

During these high school years Griffin seemed to be more of a nascent serious scientist than, say, Darwin, who had spent his undergraduate days hunting and collecting beetles rather than studying. For example, young Griffin thought he would be able to describe a new subspecies of California mice but then realized his hopes were based on errors in the literature, his first realization of the fragility of scientific fact. He tried to estimate the population of various hunted species by obtaining the number of animals killed from the state game authorities. He spent several weeks learning about bird-banding at a major banding station and was then authorized to set up a banding substation of his own near his home.

———

Soon he combined his interests in trapping small mammals and banding birds by banding bats. Recruiting friends, he banded tens of thousands of little brown bats, *Myotis lucifugus* (figure 11.1). (For the rest of his life he readily found research volunteers to help in such things as lugging heavy electronic equipment into the field, climbing into unexplored caverns, following birds in an airplane, building huts on remote sand spits, and navigating Amazon rivers in dugout canoes full of recording devices.) This bat-banding project resulted in finding that bats migrated between caves in Vermont and nurseries as far away as Cape Cod. Eventually it produced evidence of homing after displacement of more than 50 miles and of unsuspected longevity of these animals. It also yielded his first scientific publication, as a Harvard freshman, in 1934.[1]

Griffin's sailing interests led to his second paper. While sailing in the summer before entering college, he had encountered several seal carcasses left by hunters who only wanted their noses for the bounty provided by the state. Little was known about what these animals ate, so he collected the contents of their stomachs and, with the help of several curators at Harvard's Museum of Comparative Zoology, identified their contents. In one of his characteristically dry and self-effacing memoirs Griffin tells of how, many years later when he was chair of the Harvard biology department, some young discontented molecular biologist in the department sent him reprint requests for this paper in the names of several well-known molecular biologists. Griffin actually sent out the faded reprints until he realized it was a hoax.

UNDERGRADUATE YEARS

As an undergraduate biology major, Griffin took his first science courses but reported mediocre grades in everything but the courses on mammals or birds. At this time John Welsh was studying circadian rhythms in invertebrates and encouraged Griffin to do so in bats. This was an interesting problem because the bats hibernated for long periods under constant conditions in dark caves. Griffin brought some of his bats into the lab, and, using the

———

Figure 11.1

Myotis lucifugus, the little brown bat, photographed in flight by H. E. Edgerton, from the frontispiece of Griffin's *Listening in the Dark* (1958). Note the open mouth, presumably emitting high-frequency sounds.

standard physiological instrument of the time, the smoked-drum kymograph, he showed that indeed they had endogenous rhythms under constant conditions, yielding another paper in the *Journal of Mammalogy.*

Griffin knew Lazzaro Spallanzani's (1729–1799) work on bat orientation. In a brilliant series of experiments with all the requisite controls, Spallanzani had demonstrated that bats do not require their eyes but do need their ears, to navigate. He speculated that, perhaps, the sound of the bats' wings or body might be reflected from objects.[2] Griffin also was familiar with the English physiologist Hartridge's suggestion that bats might use sounds of high frequency to orientate. At this time a Harvard physics professor, G. W. Pierce, had just developed devices (the first of their kind) that could detect and produce high-frequency sounds above the human hearing range. Two fellow students, James Fisk (later president of Bell Labs) and Talbot Waterman (later a Yale zoology professor), suggested to Griffin that he take his bats to Pierce to find out whether they produced high-frequency sounds.

Pierce was quite enthusiastic about the idea. In fact he had been studying the ultrasonic sounds of insects (with the help of Vince Dethier, later the doyen of U.S. experimental entomologists). When they put the bat in front of Pierce's parabolic ultrasonic detector, they observed that the bats were producing sounds that the humans could not hear, but when the animals were flying around the room no such sounds were detected. Nor did the production of high-frequency sound seem to have any effect on the flying bats' ability to orient. When they published their observations, they suggested that the function of the supersonic sounds might be in social communication rather than orientation. (Later, Griffin realized that the detector had not been sufficiently directional to pick up the bat signals in flight. Even later, the social communication role for certain bat ultrasonic cries was confirmed.)

When Griffin was a senior, he was in a quandary about applying to Harvard's graduate school in biology because its faculty had little regard for Griffin's current interest in bird navigation. "Wiser heads emphasized that if

I really wanted to be a serious scientist I should put aside such childish interests and turn to some important subject like physiology." The problem was
solved with the announcement of the joint appointment of Karl Lashley to
the Harvard psychology and biology departments. Lashley's appointment
had been the result of the command of Harvard's president, James B.
Conant, to hire "the best psychologist in the world." Karl Lashley was the
leading "physiological psychologist" of his time and the teacher of many
subsequently famous students of brain function and behavior.[3] His particular interest to Griffin was that he had written a long and authoritative historical and experimental paper on bird homing (with J. B. Watson, later the
founder of behaviorism)[4] and had carried out his own experiments on orientation in terns. Lashley took him on as a graduate student but encouraged
him to take several courses in experimental psychology, which he did.

GRADUATE SCHOOL

In graduate school Griffin met another student, Robert Galambos (b. 1914),
who was recording cochlear microphonics from guinea pigs under Hallo-
well Davis, a leading auditory physiologist at Harvard Medical School, and
suggested Galambos look for bat cochlear microphonics in response to high-
frequency sounds. They borrowed Pierce's instruments, and Galambos was
soon able to demonstrate responses of the bat ear to ultrasonic sounds. In a
series of experiments Griffin and Galambos then showed that bats do indeed
avoid obstacles by hearing the echoes of their cries. Here is a recent recollection by Galambos of these experiments:

> Don divided a sound treated experimental room into equal parts
> by hanging a row of wires from the ceiling. We aimed the mi-
> crophone of the Pierce device at this wire array, and began to
> count the number of times a bat flying through the wires will
> hit them when normal, or deaf or mute.... The impairments
> we produced [by plugging the ears or tying the mouth shut]

were all reversible.... We also recorded the output of the Pierce device and correlated the bat's vocal output as it approached the barrier with whether it hit or missed the wires.... Everything we predicted did happen. Nothing ever went wrong. We never disagreed.... We suspected our claims might be controversial and decided a movie demonstration might help silence the skeptics.[5]

(In recent years, the original movie has been increasingly shown on nature and science TV programs in many different countries around the world.)

Needless to say, the scientific community was very skeptical at first but the film and visits to their laboratory were soon convincing. As Griffin put it later, "Radar and sonar were still highly classified developments in military technology, and the notion that bats might do anything even remotely analogous to the latest triumphs of electronic engineering struck most people as not only implausible but emotionally repugnant."

These experiments establishing bat echolocation were reported in Griffin and Galambos's two seminal papers and formed part of the latter's doctoral thesis. Griffin's thesis, by prior agreement, was on bird navigation, the problem he had originally planned to study in graduate school experiments. The central question was whether birds released in unfamiliar territory immediately determined the homeward direction and flew directly back to their nests. He captured petrels, gulls, and terns and transported them, often in rotating cages, in different directions from the site of their capture, then released them and timed their return home. However, their flight times home were consistent with both a search until they found familiar landmarks and a leisurely but direct route home.

Directly tracking them should disambiguate these possibilities, he hoped. So he got Alexander Forbes (professor of physiology at Harvard Medical School and one of the founders of modern neurophysiology) to take him up in Forbes's single-engine plane to try to track some gulls. Later, Griffin took flying lessons and bought his own two-seater with funds from the Harvard Society of Fellows. The results were again consistent with both a search pattern and true homing.

The Society of Fellows awarded three-year Junior Fellowships with generous research funds. The fellowship was originally supposed to be a super elite substitute for a Ph.D. with no required courses, teaching, exams, degrees, or requirements except for attending candlelit dinners along with the senior fellows. In practice, when the junior fellows went on the job market, say in distant Berkeley, they were told, in effect, "no degree, no job" and had to go back and get conventional doctorates. Today most junior fellows earn their doctorates first and it is a kind of fancy postdoc club, imitated predictably at such places as Princeton and Columbia. Griffin was fortunate to get elected to a Junior Fellowship, since his undergraduate grades had been too poor for a conventional graduate fellowship.

WARTIME

With the onset of war in 1941, Griffin became involved in war research at Harvard. His first assignment was to S. S. Stevens's psychoacoustic laboratory. (Stevens was the founder of modern psychophysics.) There Griffin worked on auditory communication problems and acquired valuable familiarity with acoustic equipment. After a stint in the Harvard fatigue laboratory (working on such problems as the optimal gloves for handling fly buttons) he worked with George Wald (subsequently a Nobel laureate) on problems of night vision.

One rather weird wartime incident was the Bat Bomb project.[6] Lytle Adams came to Griffin with the idea of equipping bats with small incendiary bombs and releasing them by plane over Tokyo where they would roost in Japanese "paper" houses and set fire to them. The government was supporting this idea, and Griffin agreed to help until he realized that there was no way bats could carry an adequate payload. In spite of Griffin's disavowal of its feasibility, the Bat Bomb project continued on, even involving at one point Louis Fieser, the distinguished organic chemist and inventor of napalm. In his account years later, Adams continued to defend the project and claimed that it would have ended the war in a quicker and more humane way than Hiroshima and Nagasaki.

After the war, Griffin moved to the Cornell zoology department for seven years before returning to Harvard for another twelve years. The next paragraphs summarize some of his research interests in those years.

FURTHER RESEARCH ON BAT NAVIGATION

Research on bat echolocation (Griffin's term) expanded in a number of different directions with an increasing number of collaborators.[7] (Indeed by the time of his death, most of the now numerous bat researchers everyplace in the world saw themselves directly or indirectly, implicitly or explicitly as Griffin's collaborators.) One such direction was to determine the limits of the avoidance and object-detection abilities afforded by echolocation. It was clear early that *Myotis* could discriminate wires down to a quarter of a millimeter, but could they actually echolocate moving-insect prey in the dark? Field experiments suggested that they could. This was confirmed by combining acoustic recording with ultra-high-speed strobe photography in an enclosure with released fruit flies and then weighing the bats before and after a short period of catching flies. Furthermore, the bats could quickly learn to discriminate pebbles and other inedible objects from flying insects. These experiments were carried out with Alan Grinnell, Fred Webster, and others.

Another direction initiated by Griffin, with his collaborators Alan Grinnell and Nobuo Suga (and encouraged by Galambos), was the neurophysiology of bat echolocation. Today, largely due to the work of Suga and his students, more seems to be known about the organization of auditory cortex in the bat than almost any other animal.[8]

Whereas the North American bats initially studied by Griffin emitted brief frequency-modulated (FM) signals, in 1950, F. P. Mohres discovered that the European horseshoe bat used longer-duration constant-frequency signals for echolocation. This inspired Griffin, Alvin Novick, and other collaborators to survey the signals produced by different species of bats. As most bat species are tropical, this led Griffin, Novick, and their collaborators to a number of exciting Latin American expeditions and the discovery of many different modes of echolocation, including one specialized for fishing and others in cave-dwelling birds.

In the last weeks of his life Griffin was out "night after night" on Cape Cod, "still trying to learn more about bats."

BIRDS AND OTHER CREATURES

Griffin continued to work on the mysteries of bird navigation.[9] What made this a difficult problem was that although it became clear that birds (or some birds under some conditions) were using such cues as the elevation of the sun, the pattern of the stars, the magnetism of the earth, their own circadian rhythms, and spatial memory, it was difficult to sort out the interaction and relative roles of these cues. Griffin pioneered the use of airplanes, radar, and high-altitude balloons to study this problem. (My first publication in a scientific journal, on bird navigation, arose out of a paper I wrote for an undergraduate seminar with Griffin. I then conducted research in his lab on the subject. My most vivid, if irrelevant, memory was the time he asked me to get the car battery from the next room for use as a power supply and I answered, "What does it look like?" He gave this Brooklyn boy a brief strange look and then went and picked it up himself).

Griffin's discovery of a "new sense" in bats probably influenced, at least in part, the discovery of other "new animal senses" such as infrared vision in snakes, infrasonic signals in elephants, and orientation and discrimination in electric fish. He played a more direct role in the story of the dancing language of bees. During the war the Austrian zoologist Karl von Frisch had discovered that honeybees could communicate the distance, direction, and desirability of food sources by a dancelike behavior. This work was hardly known in America in 1949 when Griffin arranged for him to give a series of lectures at Cornell, and then across the country, and shepherded their publication through Cornell University Press.[10] Griffin had initially been skeptical until he replicated some of the critical experiments himself. (At the age of 72, Griffin published his last experimental paper; it was on bees.)

Griffin was interested in how beavers communicate. The last weeks of his life found him introducing microphones into beaver's nests near the Harvard Field Station in Concord. Indeed, the number of anecdotes about the field studies he carried out in his last, and eighty-eighth, year that I collected while preparing this memoir are a measure of the man.

The Rockefeller Institute and Back to Harvard

In 1965 Griffin left Harvard to organize a new Institute for Research in Animal Behavior, jointly sponsored by Rockefeller University and the New York Zoological Society. It eventually included a field station in Millbrook, New York. Joined by the leading ethologist Peter Marler, and by Ferdinando Nottebohm, the well-known investigator of bird song and adult neurogenesis, the institute became one of the leading United States centers for the study of animal behavior. Among Griffin's collaborators and students at the institute were Roger Payne, discoverer of acoustic hunting by owls and of whale songs and now the leading advocate of whale conservation; Jim Gould, who extended von Frisch's bee studies; and Carol Ristau, pioneer in the study of intentionality in the piping plover. From 1979 to 1983 Griffin was president of the Henry Frank Guggenheim Foundation and he used this position to encourage research on animal behavior.

When Griffin retired from Rockefeller in 1986 he spent a year at Princeton University and then returned to Harvard, where he worked at the Concord Field station and occasionally taught undergraduates. In this final period of his life he continued his experimental work on bats, birds, and beavers as well as his cognitive ethology advocacy, described in the next section.

COGNITIVE ETHOLOGY

For about the first 40 years, Griffin's career had been that of the very hard-nosed empiricist and skeptic typified by the following oft-told tale (attributed to Griffin's students Donald Kennedy, former president of Stanford and FDA commissioner, and Roger Payne, among others): When passing a flock of sheep while traveling in a car, his companion noted that among the flock of sheep there were two that were black. Griffin replied, "They're black on the side facing us, anyway." Then in 1976 Griffin began to publish a series of books and papers that contained no new data, no figures, but a host of citations and arguments from philosophers as well as scientists which challenged the contemporary world view of animals. He claimed that animals (and not just chimpanzees or even mammals) were aware and conscious and these properties of their minds should be the subject of scientific study, a field he named "cognitive ethology."

At least at the beginning, these claims and exhortations were usually greeted by harsh and angry criticism (one critic called them the "satanic verses of animal cognition") or the sadness of seeing a great experimenter supposedly slipping into premature senility. (He himself even called this interest an example of "philosopause.") To better understand why imputing awareness or even minds to animals was considered outrageous, or at least, extrascientific, by most of those who studied animal behavior, we need to go back to Charles Darwin and the beginning of modern biology.

One of Darwin's central points was the continuity of humans and other animals. As evidence of mental continuity Darwin cited examples from animals of humanlike emotions of joy, affection, anger, and terror as well as of what we now call cognitive functions such as attention, memory, imagination, and reason.[11] George Romanes continued this tradition in what became known as the "anecdotal school."[12] C. Lloyd Morgan reacted against this approach and formulated what became known as Lloyd Morgan's canon,[13] essentially the application of the law of parsimony (Occam's razor) to animal behavior: "In no case may we interpret an action as the out-

come of the exercise of a higher psychological faculty, if it can be interpreted as the outcome of one which stands lower in the psychological
scale." This quickly came to imply the rejection of animal consciousness
and awareness and a wariness to impute any complex cognitive functions to
animals. This tendency was reinforced by Jacques Loeb's theory of tropisms
and the Russian school of reflexology,[14] which also downplayed or denied
consciousness in animals as well as humans. All these "objectivist" tendencies
came together in the behaviorist movement, founded by J. B. Watson.[15]
The dominant figure in behaviorism, indeed in all of U.S. psychology until
the rise of cognitive psychology, was B. F. Skinner.[16] Skinner and the other
"radical behaviorists" flatly denied the validity of the scientific study of consciousness, attention, awareness, thought, and other mental phenomena in
humans as well as other animals.

The other principal group studying animal behavior was the etholo-
gists deriving from a European zoological tradition.[17] They tended to stress
the role of innate wiring in animal behavior, in contrast to the behaviorists
who stressed the role of experience. However, they too obeyed Morgan's
canon and were generally uninterested in the role of consciousness, inten-
tion, and mental experience in animal behavior. The cognitive revolution
against behaviorism starting in the 1960s brought consciousness, attention,
and awareness back into human psychology but had left other animals still
essentially mindless and unaware.[18]

Thus Griffin's plea for studying "the question of animal awareness"
(the title of his 1976 book) was fiercely counter to the prevailing ideology
in both psychology and zoology. Griffin used a variety of arguments coming
from different directions and different fields to attack this view. One central
argument was that it was simply anti-intellectual and anti-scientific to deny
any subject an objective and experimental inquiry. A second argument was
Darwin's original one: the continuity of humans and other animals. Another
argument was that animal communication, albeit admittedly fundamentally
different from human language, might provide "a window on the animal
mind."

In his next two books, *Animal Thinking* (1984) and *Animal Minds* (1992), these arguments were amplified and supported by a Romanes-like compendium of experiments and observations that greatly enhanced, at the least, the case for animal consciousness and awareness. They included studies of tool construction and use, communication, planning, deception, blindsight, cooperative hunting, and intentionality. Two new lines of evidence came into prominence. The first were a host of neurophysiological experiments seeking mechanisms of consciousness. Since most of these were invasive, such as single-neuron recording, they could only be done in animals and thus, with all due respect to Lloyd Morgan, they assumed animals were conscious, reflecting the change in the intellectual air that Griffin had helped bring about.

The second line of new evidence, increasingly prominent in Griffin's last books and papers on cognitive ethology, came from studies done by Griffin's students—such as Jim Gould on bees or Roger Payne on whales—and by the increasing number of quasi-students, investigators who were never formally his graduate students but readily acknowledged him as their mentor. These included Dorothy Cheney and Robert Seyfarth, who detailed communicative alarm calls and deception in vervet monkeys, and Irene Pepperberg, who trained a grey parrot to answer cognitive questions in English. (Even his formal students are not readily identified as he rarely attached his name to their work).

Although many biologists and psychologists are still uneasy about Griffin's attribution of consciousness to nonhumans, particularly invertebrates, there is no question that he has radically opened up the field of animal behavior to new questions, ideas, and experiments about animal cognition. Because of his own towering achievements as a meticulous and skeptical experimental naturalist, his cogent and repeated arguments about studying the animal mind, and his support and encouragement of others, coupled with his unusual modesty and soft-spoken nature, Donald Griffin was able to effect a major revolution in what scientists do and think about the cognition of nonhuman animals.

———

NOTES

The details of Griffin's early years, his student experiences, his account of the discovery of echolocation, and his continuing work on echolocation and all the Griffin quotes are from three of Griffin's works: "Early history of research on echolocation," in R.-G. Busnel and J. F. Fish, eds., *Animal Sonar Systems* (Plenum, 1980); "Reflections of an animal naturalist" in D. A. Dewsbury, ed., *Leaders in the Study of Animal Behavior* (Bucknell University Press, 1985); and "Donald Griffin," in L. Squire, ed., *The History of Neuroscience in Autobiography*, vol. 2 (Academic Press, 1998). I would like to thank the following of Griffin's colleagues for providing additional information: Robert Galambos, Alan Grinnell, James Simmons, Roger Payne, Marc Hauser, Greg Auger, Jim Gould, and Herb Terrace. This chapter was published previously in *Biographical Memoirs of the National Academy of Sciences* (86: 1–20 [2005], "Donald R. Griffin 1915–2003").

1. Griffin, 1934.

2. Spallanzani, 1932; Galambos, 1942.

3. Gross, 1998a.

4. Watson and Lashley, 1915.

5. Galambos, 2004, personal communication. See also his memoir, Galambos, 1995.

6. Couffer, 1992.

7. Griffin, 1958.

8. E.g., Suga and Ma, 2003.

9. Griffin, 1964.

10. Frisch, 1950.

11. Darwin, 1871.

12. Romanes, 1882.

13. Morgan, 1900.

14. Loeb, 1900; Pavlov, 1929; Bechterev, 1932.

15. Watson, 1924.

16. Skinner, 1953.

17. E.g., Tinbergen, 1951.

18. Gardner, 1985.

The Genealogy of the "Grandmother Cell"

A "grandmother cell" is a hypothetical neuron that responds only to a highly complex, specific, and meaningful stimulus, such as the image of one's grandmother, that is, to a single percept or even a single concept. This chapter discusses the origin of the term, and the alternative view that complex stimuli are represented by the pattern of firing across ensembles of neurons, rather than that of a dedicated "grandmother cell."

As originally conceived, a grandmother cell was multimodal, but the term came to be used mostly for representing a visual percept. As we shall see, the term arose because the first such neuron was postulated to represent a grandmother (actually, the very first was a specific mother). There might be many grandmother cells responding to a specific stimulus, such as one grandmother, but their response properties would be the same. Thus "coding by grandmother cells" is at the other extreme from "ensemble," "coarse," or "population" coding, in which a grandmother or other stimulus is coded by the pattern of activity over a group of neurons. In ensemble coding each member of the ensemble responds somewhat differently, for example to a granny's wrinkles, to white hair, or to several different old women; the coding of a specific grandmother is done by a unique pattern of activation across the ensemble.

Starting in the early 1970s the term "grandmother cell" moved from laboratory jargon and jokes into neuroscience journals and serious discussions of the bases of pattern perception.[1] The term is now nearly ubiquitous in introductory neuroscience and vision textbooks, where it often plays the role of straw man or foil for a discussion of ensemble or coarse coding theories of sensory representation.[2] This chapter considers the origins of the term "grandmother cell" and similar expressions and, more briefly, the roots of ideas about ensemble coding.

JERRY LETTVIN AND THE BIRTH OF MOTHER AND GRANDMOTHER CELLS

Jerry Lettvin originated the term "grandmother cell" around 1969 in his MIT course on "Biological Foundations for Perception and Knowledge." When discussing the problem of how neurons can represent individual objects, he told a (tall) tale of how the neurosurgeon A. Akakhievitch had located a group of brain cells that "responded uniquely only to a mother ...whether animate or stuffed, seen from before or behind, upside down or on a diagonal or offered by caricature, photograph or abstraction."[3] At this point the mother-obsessed character from Philip Roth's novel *Portnoy's Complaint*[4] appeared and Akakhievitch ablated all of the mother cells in Portnoy's brain. As a result, Portnoy completely lost the concept of his mother. (See box 12.1.) Akakhievitch then went on to the study of "grandmother cells."

From this origin, the term *grandmother cell* seems to have spread so quickly that Horace Barlow, in his 1972 paper "Single units and sensation: A neuron doctrine for perceptual psychology,"[5] didn't even explicitly define the term in criticizing the idea, and in 1973 Colin Blakemore could write of the "great debate [that] has become known as the question of the 'grandmother cell.' Do you really have a certain nerve cell for recognizing the concatenation of features representing your grandmother?"[6]

Box 12.1
Lettvin's Story about Mother and Grandmother Cells (ca. 1969)

In the distant Ural mountains lives my second cousin, Akakhi Akakhievitch, a great if unknown neurosurgeon. Convinced that ideas are contained in specific cells, he had decided to find those concerned with a most primitive and ubiquitous substance—mother.... And he located some 18,000 neurons that responded uniquely only to a mother, however displayed, whether animate or stuffed, seen from before or behind, upside down or on a diagonal, or offered by caricature, photograph, or abstraction.

He had put the mass of data together and was preparing his paper, anticipating a Nobel prize, when into his office staggered Portnoy, world-renowned for his Complaint (Roth, 1969). On hearing Portnoy's story, he rubbed his hands with delight and led Portnoy to the operating table, assuring the mother-ridden schlep that shortly he would be rid of his problem.

With great precision he ablated every one of the several thousand separate neurons and waited for Portnoy to recover. We must now conceive the interview in the recovery room.
"Portnoy?"
"Yeah."
"You remember your mother?"
"Huh?"
(Akakhi Akakhievitch can scarcely restrain himself. Dare he take Portnoy with him to Stockholm?)
"You remember your father?"
"Oh, sure."
"Who was your father married to?"
(Portnoy looks blank)
"You remember a red dress that walked around the house with slippers under it?"
"Oh, certainly."
"So who wore it?"
(Blank)
"You remember the blintzes you loved to eat every Thursday night?"
"They were wonderful."
"So who cooked them?"
(Blank)
"You remember being screamed at for dallying with shikses?"
"God, that was awful."
"So who did the screaming?"
(Blank)

Box 12.1
(continued)

And so it went.... It made no difference—Portnoy had no mother. "Mother" he could conceive—it was generic. "My mother" he could not—it was specific....

Akakhievitch then ... went back to ... "grandmother cells."

This parable is abridged from a letter Lettvin sent Horace Barlow in 1995 (described in Barlow, 1995). Much earlier, Barlow (1953) had described cells in the frog's retina as "bug detectors," but little notice had been taken. In the late 1950s Lettvin and his colleagues at MIT were studying these and other complex cells in the frog, but again the mainstream neuroscience community had ignored their work. Thus at a 1959 meeting on sensory communication at MIT, Barlow (respectable for other reasons by then) but not Lettvin had been invited; Barlow arranged for some of the participants to see experiments in Lettvin's lab. Subsequently a paper by Lettvin and his colleagues was added to the end of the meeting's Proceedings (Lettvin et al., 1961) and eventually "What the frog's eye tells the frog's brain" (Lettvin et al., 1959) became well known. Presumably, Lettvin's previous research on how the frog's retina codes complex stimuli was related to his story about mother and grandmother cells.

JERZY KONORSKI'S GNOSTIC UNITS

Although unknown to Lettvin, the grandmother cell idea had actually been set out in detail as a serious scientific proposal a few years earlier by the Polish neurophysiologist and neuropsychologist Jerzy Konorski in his *Integrative Activity of the Brain*,[7] a wide-ranging set of speculations on the neurophysiology of perception and learning (see figure 12.1 and box 12.2). His ideas on the organization of the cerebral cortex in perception anticipated subsequent discoveries to an amazing degree. Konorski predicted the existence of single neurons sensitive to complex stimuli such as faces, hands, emotional expressions, animate objects, locations, and so on (see figure 12.2). He called them "gnostic" neurons, and they were virtually identical to what later were called "grandmother cells." He suggested that the gnostic neurons

Figure 12.1
Jerzy Konorski in front of the Nencki Institute of Experimental Biology in Warsaw.
Photograph 1961 by the author.

Box 12.2
Jerzy Konorski (1903–1973)

Konorski's speculations about gnostic cells came at the end of a long and distinguished career studying the brain and behavior (Konorski, 1967, 1974; Fonberg, 1974). As medical students in Warsaw he and Stefan Miller discovered that there was another type of conditioned reflex other than the one discovered by Pavlov, namely one under the control of reward. They called it Type II to distinguish it from Pavlov's, which they called Type I. Subsequently and independently, Skinner made this same distinction and Konorski and Miller's Type II conditioning became known as operant or instrumental conditioning.

After a few years as a psychiatrist, Konorski spent two years in Pavlov's laboratory in Leningrad but never convinced the master that there really were two types of conditioned reflexes. Konorski then returned to Warsaw and set up a conditioning laboratory in the Nencki Institute of Experimental Biology. He also married and collaborated with Dr. Liliana Lubinska, who had studied neurophysiology in Paris, and through her became familiar with Western and particularly Sherringtonian neurophysiology. When the war started, Konorski was extraordinarily fortunate to be able to escape Poland. (His colleague Miller committed suicide when the Nazis arrived.) His Russian friends got him appointed head of the famous primate laboratory at Sukhumi on the Black Sea. The laboratory eventually moved to Tbilisi as the Germans approached. It was still near the front, and Konorski had a great deal of experience treating head wounds in a nearby army hospital. At the end of the war he returned to Poland and played a major role in reconstructing Polish neuroscience as head of the Department of Neurophysiology at the Nencki Institute.

In 1948 Cambridge University Press published his *Conditioned Reflexes and Neuron Organization*, which was an attempt to bring Pavlovian reflexology in line with Sherringtonian neurophysiology. In the 1960s there were close collaborative relations between Konorski's laboratory and researchers at the Laboratory of Neuropsychology at the National Institutes of Health (Hal Rosvold, Mort Mishkin, and Patricia Goldman). In addition to the concepts of gnostic units and gnostic fields, Konorski's *Integrative Activity of the Brain* (1967) contains many important and influential ideas about learning and memory.

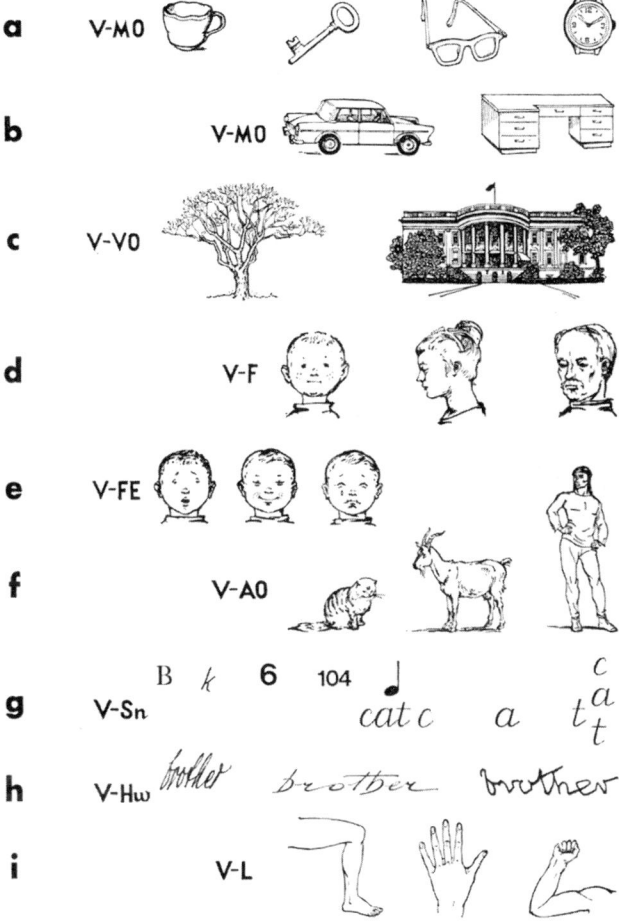

Figure 12.2
This illustration is taken from Konorski, 1967, figure III-1, "Particular categories of visual stimulus-objects probably represented in different gnostic fields." (a) Small manipulable objects; (b) larger partially manipulable objects; (c) nonmanipulable objects; (d) human faces; (e) emotional facial expressions; (f) animated objects; (g) signs; (h) handwriting; (i) positions of limbs. Used by permission of University of Chicago Press.

were organized into specific areas of the cerebral cortex he termed "gnostic fields." That is, he predicted (correctly in many cases) the existence of areas of the cortex devoted to the representations of such things as faces, emotional expressions, places, and spatial relations. Destruction of a gnostic field would lead, he predicted, to what were later described as category-specific agnosias. Furthermore, he localized many of these gnostic fields, such as the face field in ventral temporal cortex and the space field in posterior parietal cortex (figure 12.3). Overall, these gnostic fields and their locations are remarkably similar to contemporary views of the putative functions of extrastriate cortex based on monkey single-neuron studies and human imaging experiments.[8]

At the time of their publication, there was nothing in the literature anything like Konorski's ideas of highly specialized perceptual neurons in mammals or of areas of the cortex devoted to the representations of particular classes of visual stimuli. In retrospect, however, it is possible to delineate the origins of Konorski's speculations.

Konorski's views of the neural organization of perception were a synthesis and extension of three lines of work in the decade before the publication of his book. The first was Hubel and Wiesel's demonstration of the hierarchical processing of sensory information: how as one proceeds from center-surround to simple receptive fields and then to complex and then the (now revised) hypercomplex ones the selectivity of the cells increases and their ability to generalize across the retina increases.[9] The possibility that this hierarchy of increasing stimulus specificity continues beyond visual areas V2 and V3 was made explicit in their 1965 paper, which is repeatedly cited by Konorski. That paper ends as follows:

> How far such analysis can be carried is anyone's guess, but it is clear that the transformations occurring in these three cortical areas [V1, V2 and V3] go only a short way toward accounting for the perception of shapes encountered in everyday life.[10]

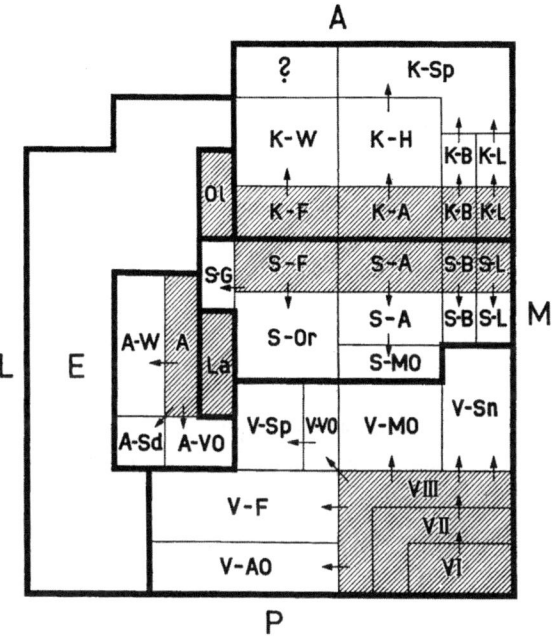

Figure 12.3

From Konorski, 1967. "Conceptual map of the human cerebral cortex." A, anterior; P, posterior; L, lateral; M, medial. Projective fields are hatched; gnostic fields are plain. The modality boundaries are thick lines. The arrows denote connections. The numbers in parentheses are tentative correspondences with Brodmann's areas. The letters signify the gnostic fields shown in figure 12.2. V, visual analyzer: V-I (Brodmann's area 17); V-II (18); V-III (19); V-Sn, sign visual field (7b); V-MO, field for small manipulable objects (7b); V-VO, field for large objects (39); V-Sp, field for spatial relations (39, right hemisphere); V-F, field for faces (37); V-AO, field for animated objects (37). A, auditory analyzer: A, projective auditory field (41, 42); A-W, audio-verbal field (22); A-Sd, field for various sounds (22, right hemisphere); A-VO, field for human voices (21). The legends for the symbols for the somesthetic (S) and kinesthetic (K) fields have been omitted. Ol, olfactory analyzer; E, emotional analyzer. Used by permission of University of Chicago Press.

A second line of inspiration for Konorski's ideas was the research by Karl Pribram and his students, particularly Mort Mishkin, on the cognitive effects of lesions on what was then called "association cortex" in monkeys.[11] From his close association with Hal Rosvold and Mishkin (both commented on earlier drafts of his book) Konorski was well aware that lesions of inferior temporal (IT) cortex produced specific impairments in visual cognition in monkeys and that similar areas of "association cortex" existed for audition and somesthesis. Today, at the annual meeting of the Society for Neuroscience, there are multiple sessions on IT cortex under the general rubric of "Vision." However, at that time most visual neurophysiologists had never heard of this area and did not realize that it had visual functions, let alone that it sat at the top of a series of hierarchically arranged extrastriate visual areas. Indeed, although V2 and V3 had been described, no other extrastriate visual areas such as MT or V4 were known until 1971.[12] Citing Pribram and Mishkin,[13] Konorski wrote:

> In monkeys the gnostic visual area seems to be localized in inferotemporal cortex, as judged from numerous experimental results in which ablations of this region produced impairment of visual discrimination.[14]

The third line of evidence for his theories of gnostic neurons and areas came from Konorski's familiarity with the various agnosias that follow cortical lesions in humans from his own clinical experience, from the Western neuropsychological literature, and from Luria's work in the Soviet Union. He was aware of both the symptomatic specificity of some cases of agnosia and their tendency to be localizable. Furthermore, unlike most contemporary neuropsychologists and neurophysiologists he was aware of the similarity of human agnosias to the effects of experimental lesions in monkeys. For example, he directly related prosopagnosia, or face agnosia, after ventral temporal lesions in humans to the visual learning deficits in monkeys after inferior temporal lesions.

In summary, Konorski's prophetic ideas on gnostic neurons and gnostic fields came from a bold extension of Hubel and Weisel's findings to account for specific cognitive effects of specific lesions in monkeys and humans.

Konorski's book received a long and laudatory review in *Science* (by me).[15] However, for at least the next decade virtually all of the many citations to the book were to the parts concerned with learning rather than perception; learning theory still dominated American psychology. As described in the next section, the ideas on gnostic neurons did influence one laboratory, namely mine, the laboratory that first reported (the predicted) neurons in IT cortex that selectively respond to faces and hands.

In the last decade gnostic cells have begun to be commonly mentioned in textbooks and in the vision and pattern-recognition literature, usually as synonyms for grandmother cells and usually in the context of inferior temporal cortex cells.

The Discovery of Face- and Hand-Selective Cells in Inferior Temporal Cortex of the Monkey

In the late 1960s my colleagues and I reported visual neurons in inferior temporal cortex of the monkey that fired selectively to hands and faces.[16] These observations were probably primed by our familiarity with Konorski's gnostic units as well as the propinquity of Lettvin's work on detectors in the frog's eye,[17] Hubel and Wiesel's discoveries on the hierarchical processing in cats and monkeys, and local talk about grandmother cells. Starting about twelve years later, these findings were replicated and extended in a number of laboratories[18] and were often viewed as evidence for grandmother cells. Konorski himself saw them as confirming his ideas of gnostic cells.[19] For some time these cells were the strongest evidence for the existence of grandmother/gnostic cells. However, there has been no good evidence for cells from monkeys that are selective for other visual objects important or common for monkeys such as fruit, tree branches, monkey genitalia, or other features in their natural environments. However, inferior temporal cells can be trained to show great specificity for arbitrary visual objects and these would seem to fit the requirements of gnostic/grandmother cells.[20] Furthermore, there is now good evidence for cells in the human hippocampus that have highly selective responses to gnostic categories, including highly selective responses to individual human faces.[21]

However, most of the reported face-selective cells do not really fit a very strict criteria of grandmother/gnostic cells in representing a specific percept, that is, a cell narrowly selective for one face and only one face across transformations of size, orientation, and color.[22] Even the most selective face cells usually also discharge, if more weakly, to a variety of individual faces. Furthermore, face-selective cells often vary in their responsiveness to different aspects of faces, suggesting that they form ensembles for the coarse or distributed coding of faces rather than detectors for specific faces. Thus, a specific grandmother may be represented by a specialized ensemble of grandmother or near-grandmother cells.

There are two reasons why the members of face-coding ensembles may appear more specialized than the members of other stimulus-encoding ensembles, that is, why there are many more face cells than banana cells. First, it is more crucial for a monkey (or human) to differentiate among faces than among any other categories of stimuli such as bananas. Second, faces are more similar to each other in their overall organization and fine detail than any other stimuli that a monkey must discriminate among. If there had been strong selective pressure for a monkey to distinguish individual bananas, it would probably have ensembles for doing so that were made up of cells selective for banana in general, but which showed graded response to different characteristics of bananas.

LABELED LINES AND HIERARCHIES

Two central characteristics of grandmother/gnostic cells have a long history. The first is that they are examples of labeled line coding, and the second is that they are at the top of a hierarchy of increasing convergence.

Labeled line coding refers to activity in a neuron coding a particular stimulus property, such as red, or a grandmother. This specificity derives from the connections of the neuron, not from the pattern of the neuron's firing as is the case for various temporal codes such as rate, latency, or phase locking. Perhaps the earliest notion of a labeled line was in Galen's distinction between sensory and motor nerves in the second century[23] (see chapter 2). On the basis of his experience as physician to the gladiatorial school in Pergamon he realized that section of some nerves resulted in a specific sensory loss and section of others resulted in motor loss. He thought the distinction derived from the connections of the nerves to specific regions of the brain.

The first modern labeled line theory of vision was Thomas Young's trichromatic theory of color.[24] Johannes Müller (1838) then generalized this idea to all the senses in his doctrine of specific nerve energies.[25] In that doctrine, when a given nerve type (or nerve energy in his terms) is excited, the same type of experience is produced independent of the cause of the excitation. The first example of labeled line coding by single-neuron activity was probably Adrian and Matthews' finding that action potentials in a given optic nerve fiber of the conger eel signaled the photic stimulation of a specific part of the eel's retina.[26]

Turning to the other property of grandmother cells, convergence, the most extreme example of neural convergence is William James's concept of a "pontifical cell" whose activity is identical to consciousness, as in this passage from his *Principles of Psychology*:

> There is, however among the cells one central or pontifical one
> to which our consciousness is attached. But the events of all the
> other cells physically influence this arch-cell; and through pro-
> ducing their joint effects may be said to "combine."[27]

C. S. Sherrington in his classic *Man on his Nature* took up James's ideas that there might be "convergence ... of the nervous system ... onto one ultimate "pontifical nerve-cell."[28] He then rejected pontifical cells in favor of an ensemble cell theory of consciousness as "a million-fold democracy whose each unit is a cell."

Barlow thought that the proposal of grandmother cells was inadequate because the multidimensional aspects of visual percepts could not be represented by a single individual cell.[29] Rather, he proposed that a small number of cells would be needed to represent a percept. He named such cells "cardinal" cells since cardinals are lower in the hierarchy than popes and there are more of them.

CONCLUDING COMMENT

The idea that there might be convergence of neural input onto a single cell which would provide that cell with the ability to represent a complex and specific percept seems to have arisen independently several times, first as the gnostic cells elaborated in detail by Konorski and then as the grandmother cells deriving from Lettvin's parable. Cells with properties that are similar to those of gnostic and grandmother cells have been found in both the inferior temporal cortex and the hippocampus. Grandmothers (and other complex objects) may be represented by ensembles of "grandmother" cells that vary in their responses to different aspects of the stimulus. Finally, contemporary human brain imaging studies have yielded specialized regions of the cortex that closely resemble the gnostic fields proposed by Konorski.

POSTSCRIPT

Since our discovery of neurons in monkeys that respond much more strongly to images of faces than to other images or to scrambled faces[30] there has been an enormous increase in our understanding of these neurons. We now know that some of these neurons are more sensitive to facial identity, some to emotional expression of faces,[31] and some to the gaze position of the eyes.[32]

FACE-SELECTIVE NEURONS

Face-selective neurons are found throughout inferior temporal cortex but are concentrated in two regions of the superior temporal sulcus, located on the dorsal border of the inferior temporal (IT) cortex.[33] They are also found immediately dorsal to IT (in the superior temporal polysensory area), in the amygdala, and in the ventral prefrontal cortex.[34] The face-selective cells in these different areas are thought to play different roles in the processing of the facial image.

Face-selective cells may be selective for face orientation or may generalize across different lateral orientations.[35] The response to inverted faces is much reduced.[36] As is usually the case with other IT neurons, their response selectivity remains invariant over changes in size, contrast, and exact retinal location.[37]

Furthermore, face-selective cells, at least in IT cortex, do more than process the retinal image; they have other cognitive functions as well. Like other IT neurons, they can be strongly modulated by attention and have mnemonic properties in both short and long time scales.[38] Although probably present at birth, they can be modified with experience.[39]

Overall these properties of face-selective cells suggest strongly that they may be involved in the behavioral detection and discrimination of faces. This idea is confirmed by the demonstration that electrical stimulation of a cluster of IT cells elicits face recognition.[40]

Two early generalizations made in the chapter remain valid. The first is that such apparent gnostic neurons have been found only for faces and body parts and not for other natural stimuli important for a monkey. The second is that face-selective cells do not fit strict criteria of "grandmother cells." Rather than single, dedicated grandmother cells, an ensemble of face-selective cells is required to represent a specific face.

FACE PROCESSING IN HUMANS

Whereas it took about 12 years for anyone to try and replicate our basic findings on IT neurons in monkeys, it took about 19 years for the search for responses to faces in the temporal cortex of humans to begin. Selective responses to faces were finally shown in human ventral temporal cortex, especially the fusiform gyrus, by PET scanning, single-unit recording, field potentials, and fMRI.[41] Since then there has been a flood of imaging papers on face processing in the human temporal lobe.

Functional magnetic resonance imaging (fMRI) has revealed three main face-selective patches in humans: the occipital face area (OFA), the superior temporal face area (STS-FA), and the most studied and first discovered fusiform face area (FFA). The OFA has been thought to be involved in processing face components; the FFA in processing face identity; and the STS-FA in processing gaze direction.[42] Recently up to two anterior face-selective regions have been described, perhaps involved in face recognition.

There are two current questions about the FFA. The first is whether the FFA is a "pure" face area or an "expertise" area; the current evidence suggests that within the FFA there are exclusive face-processing mechanisms. The second question is on the role of the widespread distributed processing of the facial image throughout human ventral cortex outside of the three "focal" areas: what are the relative roles of the focal and distributed processing of faces?

MONKEY NEURONS AND HUMAN FACE-SELECTIVE PATCHES

In a beginning attempt to bridge monkey single-unit and human fMRI studies, we and others have used fMRI to demonstrate several face-selective patches in monkey temporal cortex.[43] Tsao and her colleagues showed that in a face-sensitive patch (as detected using fMRI) in the monkey that may correspond to human FFA, virtually all the single neurons recorded were face selective.[44] Subsequent fMRI studies showed that there were at least five selective face patches in the monkey temporal lobe,[45] the largest two corresponding to the areas in which previous researchers had found the highest concentration of face-selective cells. Electrical stimulation of four of the face patches revealed by fMRI showed that they were interconnected, forming a "specifically interconnected hierarchical network."[46] Furthermore, both Tsao et al. and Pinsk et al. showed that the multiple face patches in monkeys corresponded closely to the multiple face patches in humans.[47] Beyond beginning to establish homologies between humans and monkeys, further elucidation of the temporal face-processing network is likely to throw light on the circuitry underlying not only face processing but also other types of pattern recognition.

A Candidate for True Grandmother Cells?

As noted above, it was generally agreed that inferior temporal cells coded faces by small ensembles of cells and rather than by dedicated "grandmother cells."[48] Similarly, there were an increasing number of examples from other systems where a small number of cells (or "sparse coding") were sufficient to code complex phenomena, such as place cells in rat hippocampus[49] or time in HVC neurons of the song system in songbirds.[50]

Then in 2005 a group at Cal Tech reported cells in the human medial temporal lobe (hippocampus and several adjacent areas) that responded with astonishing invariance and specificity.[51] For example, one responded only or best to various pictures of the actress Halle Berry including images of her dressed as a "cat woman" in a screen role of hers. "Notably," to quote the authors, "it was selectively activated by the letter string 'Halle Berry'." The cell did not respond to other faces, cat women, or letter strings (see figure 12.4). Another cell responded best to various pictures of the Sydney opera house and the letter string "opera house" but not to a variety of other buildings and letter strings. Both Ms. Berry and the opera house were familiar to the subject, and these stimuli were chosen for further analysis because in preliminary testing they had elicited responses. These results were

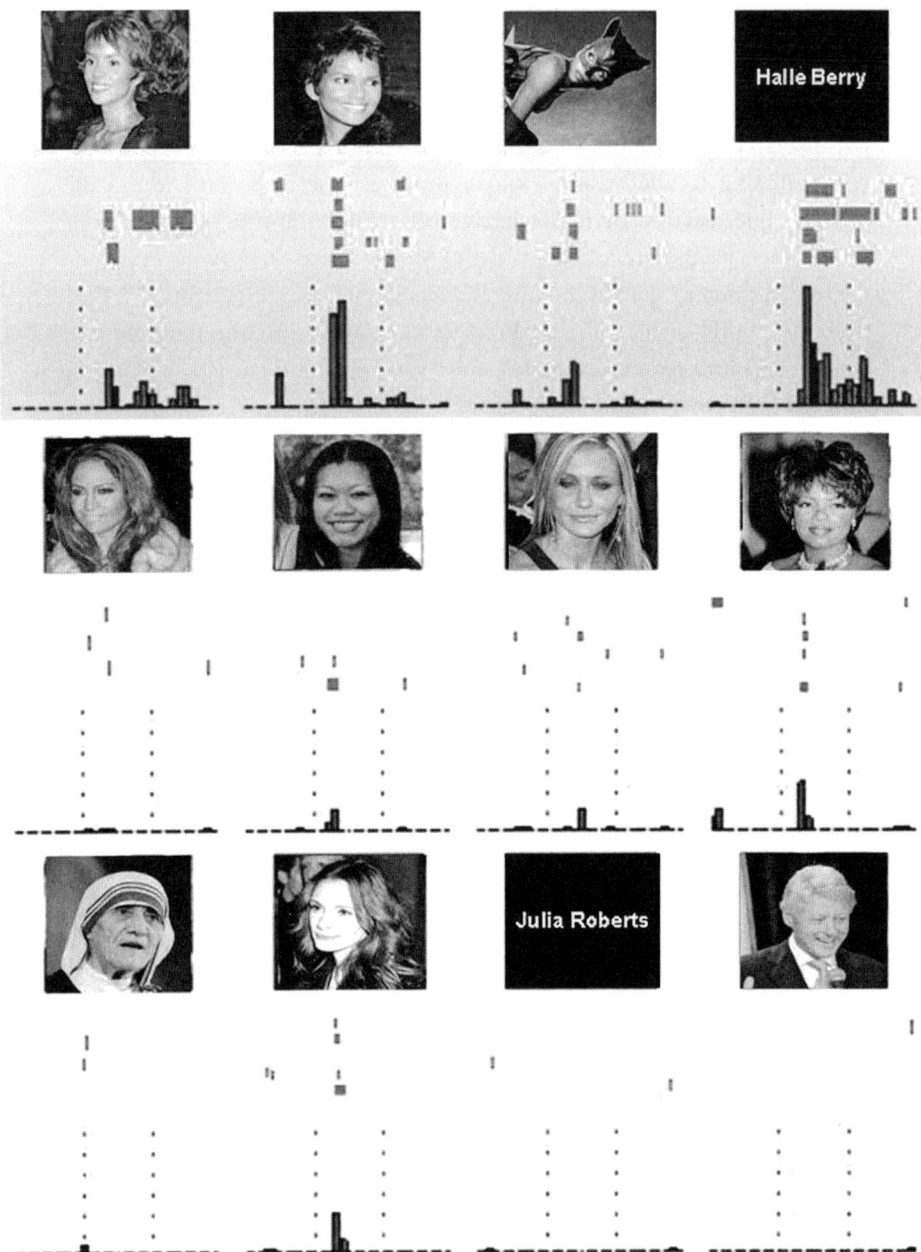

Figure 12.4

A single unit in the right anterior hippocampus activated by different views of the actress Halle Berry and not by a variety of other faces. The cell also responds to a drawing of her and to herself dressed as Catwoman in a recent movie and to the letter string of her name. The vertical dashed lines mark stimulus onset and offset, which were 1 second apart (Quiroga et al., 2005). Reprinted by permission from Macmillan Publishers Ltd.

certainly much closer to illustrating the idea of a grandmother cell than any previous ones.

In a subsequent paper the authors make it clear that, for several reasons, they had found "sparse but not 'Grandmother-cell' coding."[52] First, it was simply improbable that a single cell would show invariant responses to only one individual, especially as the stimulus set tested was small. Second, in fact some cells responded to two individuals (on the same TV program) or to two sites (seen on a recent trip). Third, the usual response latency (250–350 ms) was much longer than typical recognition times. The second and third points suggested that these cells might play a role in memory for abstract concepts rather than in sensory processing.

NOTES

This chapter is adapted from an article published in *The Neuroscientist* (8: 512–518 [2002], "The genealogy of the 'grandmother cell'").

1. Barlow, 1972; Blakemore, 1973b; Anstis, 1975; Frisby, 1980; Marr, 1982; Churchland, 1986.

2. Rosenzweig et al., 1999; Gazzaniga et al., 1998; Cowey, 1994.

3. Lettvin, personal communication in Barlow, 1995.

4. Roth, 1969

5. Barlow, 1972

6. Blakemore, 1973b.

7. Konorski, 1967

8. E.g., Martin et al., 2000; Caramazza, 2000.

9. Hubel and Wiesel, 1962, 1965.

10. Hubel and Wiesel, 1965.

11. Mishkin, 1966.

12. Allman and Kaas, 1971.

13. Pribram and Mishkin, 1955.

14. Konorski, 1967.

15. Gross, 1968.

16. Gross, Bender, and Rocha-Miranda, 1969; Gross, Rocha-Miranda, and Bender, 1972; Gross, 1994.

17. Lettvin et al, 1959, 1961.

18. Perrett, Rolls, and Caan, 1982; Rolls, 1984; Yamane, Kaji, and Kawano, 1988.

19. Fonberg, 1974; Konorski, 1974.

20. Logothetis and Sheinberg, 1996; Tanaka, 1996.

21. Gross, 2000b; Kreiman, Koch, and Fried, 2000, 2001.

22. Desimone, 1991; Gross, 1992.

23. Gross, 1998a.

24. Boring, 1950.

25. Müller, 1965.

26. Adrian and Matthews, 1927.

27. James, 1890.

28. Sherrington, 1940.

29. Barlow, 1972.

30. Gross, Rocha-Miranda, and Bender, 1972.

31. Hasselmo, Rolls, and Baylis, 1989.

32. Perrett, Rolls, and Caan, 1982

33. Tsao and Livingstone, 2008, figure 8.

34. Bruce and Gross, 1981 (dorsal to IT); Leonard, Rolls, and Baylis, 1985 (amygdala); Scalaidhe et al., 1999 (ventral prefrontal cortex).

35. Desimone, Albright, and Gross, 1984.

36. Perrett et al., 1985.

37. Desimone, Albright, and Gross, 1984; Schwartz et al., 1983.

38. E.g., Miyashita, 1988, Miller, Li, and Desimone, 1991; Colombo and Gross, 1994; and Desimone, 1996.

39. Rodman, O'Scalaidhe, and Gross, 1993; Logothetis, Pauls, and Poggio, 1995.

40. Afraz, Kiani, and Esteky, 2006.

41. Haxby et al., 1991; Sergent and Signoret, 1992; Ojemann, Ojemann, and Lettich, 1992; Allison et al., 1994; Kanwisher, McDermott, and Chun, 1997; Puce, Gore, and McCarthy, 1995.

42. See references in the review by Tsao and Livingstone, 2008.

43. Tsao et al., 2003; Pinsk et al., 2005; Logothetis et al., 1999.

44. Tsao, Freiwald, and Tootell, 2006.

45. Tsao, Moeller and Freiwald, 2008; Pinsk et al., 2009

46. Moeller, Freiwald, and Tsao, 2008.

47. Tsao, Moeller and Freiwald, 2008; Pinsk et al., 2009

48. Desimone, 1991; Gross, 1992; Abbott, Rolls, and Tovee, 1996; Rolls and Tovee, 1995.

49. Jung and McNaughton, 1993; Thompson and Best, 1989.

50. Hahnloser, Koshevnikov, and Fee, 2002.

51. Quiroga et al., 2005.

52. Quiroga et al., 2007.

———

REFERENCES

Abbott, L. F., Rolls, E. T., and Tovee, M. J., 1996. Representational capacity of face coding in monkeys. *Cereb. Cortex*, 6: 498–505.

Abrous, D. N., Koehl, M., and Le Moal, M., 2005. Adult neurogenesis: From precursors to network and physiology. *Physiol. Rev.*, 85: 523–569.

Adair, H., and Bartley, S. H., 1958. Nearness as a function of lateral orientation in pictures. *Percept. Mot. Skills*, 8: 135–141.

Adams, K. R., 1969. Reflections and laterality. *Leonardo*, 2: 301–307.

Adrian, E. D., and Matthews, B., 1927. The action of light on the eye. Part 1. The discharge of impulses in the optic nerve and its relation to the electric changes in the retina. *J. Physiol.*, 97: 378–414.

Afraz, S. R., Kiani, R., and Esteky, H., 2006. Microstimulation of inferotemporal cortex influences face categorization. *Nature*, 442: 692–695.

Alhazen, 1989 [ca. 1021]. *The Optics of Ibn al-Haytham. Books I–III: On Direct Vision*. A. I. Sabra, trans. Warburg Institute, London.

Alhazen, 2001 [ca. 1021]. *Alhacen's Theory of Visual Perception: A Critical Edition, with English Translation and Commentary, of the First Three Books of Alhacen's "De Aspectibus," the Medieval Latin Version of Ibn al-Haytham's "Kitab al-Manazir."* A. M. Smith, trans., ed. American Philosophical Society, Philadelphia.

Alhazen, 2006 [ca. 1021]. A. M. Smith, trans., ed. *Alhacen on the Principles of Reflection: A Critical Edition, with English Translation and Commentary, of Books 4 and 5 of Alhacen's "De Aspectibus," the Medieval Latin Version of Ibn al-Haytham's "Kitab al-Manazir."* American Philosophical Society, Philadelphia.

Allen, E., 1912. The cessation of mitosis in the central nervous system of the albino rat. *J. Comp. Neurol.*, 19: 547–568.

Allison, T., Ginter, H., McCarthy G., Nobre, A. C., Puce, A., Luby, M., and Spenser, D. D., 1994. Face recognition in human extrastriate cortex. *J. Neurophysiol.*, 71: 821–825.

Allman, J. M., and Kaas, J. H., 1971. A representation of the visual field in the caudal third of the middle temporal gyrus of the owl monkey (*Aotus trivirgatus*). *Brain Res.*, 31: 85–101.

Alting, M.P.C. and Waterbolk, T. W., 1982. New light on the anatomical errors in Rembrandt's *Anatomy Lesson of Dr. Nicolaas Tulp. J. Hand Surg.* 7: 632–634.

Altman, J., 1962. Are new neurons formed in the brains of adult mammals? *Science*, 135: 1127–1128.

Altman, J., 1963. Autoradiographic investigation of cell proliferation in the brains of rats and cats. *Anat. Rec.*, 145: 573–591.

Altman, J., 1967. Postnatal growth and differentiation of the mammalian brain, with implications for a morphological theory of memory In: Quarton, G. C., Melnechuck, T., and Schmitt, F. O., eds., *The Neurosciences: A Study Program.* Rockefeller University Press, New York.

Altman, J., 1969. Autoradiographic and histological studies of postnatal neurogenesis. IV. Cell proliferation and migration in the anterior forebrain, with special reference to persisting neurogenesis in the olfactory bulb. *J. Comp. Neurol.*, 137: 433–458.

Altman, J., 2009. The discovery of adult mammalian neurogenesis. In: Seki, T., ed. *Neurogenesis in the Adult Brain.* Springer, Berlin.

Altman, J., and Bayer, S. A., 1995. *Atlas of Prenatal Rat Brain Development.* CRC Press, Boca Raton, FL.

Altman, J., and Bayer, S. A., 1996. *Development of the Cerebellar System: In Relation to Its Evolution, Structure, and Functions.* CRC Press, Boca Raton, FL.

Altman, J., and Das, G. D., 1965. Autoradiographic and histological evidence of postnatal hippocampal neurogenesis in rats. *J. Comp. Neurol.*, 124: 319–335.

Altman, J., and Das, G. D., 1966a. Autoradiographic and histological studies of postnatal neurogenesis. I. A longitudinal investigation of the kinetics, migration and transformation of cells incorporating tritiated thymidine in neonate rats, with special reference to postnatal neurogenesis in some brain regions. *J. Comp. Neurol.*, 126: 337–390.

Altman, J., and Das, G. D., 1966b. Autoradiographic and histological studies of postnatal neurogenesis. II. A longitudinal investigation of the kinetics, migration and transformation of cells incorporating tritiated thymidine in infant rats, with special reference to postnatal neurogenesis in some brain regions. *J. Comp. Neurol.*, 128: 431–474.

Anderson, P., Morris, R., Amaral, D., Bliss, T., and O'Keefe, T., eds., 2006. *The Hippocampus Book*. Oxford University Press, Oxford.

Anonymous, 1881a. The festivities of the congress. *Br. Med. J.*, 2: 303–304, 752, 822–824, 836–842.

Anonymous, 1881b. The charge against Professor Ferrier under the Vivisection Act: Dismissal of the summons. *Br. Med. J.*, 2: 836–842.

Anstis, S. M., 1975. What does visual perception tell us about visual coding? In: Gazzaniga, M. S., and Blakemore, C., eds., *Handbook of Psychobiology*. Academic Press, New York.

Aretaeus, 1856 [2nd C.]. The book of cures. In: Adams, F., ed. and trans., *The Extant Works of Aretaeus, the Cappadocian*. Sydenham Society, London.

Arnheim, R., 1974. *Art and Visual Perception*. University of California Press, Berkeley.

Arnott, R., Finger, S., and Smith, C. U. M., eds., 2003. *Trepanation: History, Discovery, Theory*. Swets and Zeitlinger, Lisse, Netherlands.

Asanuma, H., 1975. Recent developments in the study of the columnar arrangement of neurons within the motor cortex. *Physiol. Rev.*, 55: 143–156.

Avicenna [Ibn Sina], 1930 [11th C.]. *The Canon of Medicine*. O. C. Gruner, trans. Luzac, London.

Bakay, L., 1985. *An Early History of Craniotomy*. Charles C. Thomas, Springfield, IL.

Bal, M., 1991. *Reading "Rembrandt": Beyond the Word-Image Opposition*. Cambridge University Press, Cambridge.

Bango Torviso, I. G., and Marias, F., 1982. *Bosch: Reality, Symbol and Fantasy*. Silex, Madrid.

Barami, K., Iversen, K., Furneaux, H., and Goldman, S. A., 1995. Hu protein as an early marker of neuronal phenotypic differentiation by subependymal zone cells of the adult songbird forebrain. *J. Neurobiol.*, 28: 82–101.

Barber, B., 1961. Resistance by scientists to scientific discovery. *Science*, 134: 596–602.

Barcroft, J., 1932. La fixité du milieu interieur est la condition de la vie libre (Claude Bernard). *Biol. Rev.*, 7: 24–87.

Barlow, H. B., 1953. Summation and inhibition in the frog's retina. *J. Physiol.*, 119: 69–88.

Barlow, H. B., 1972. Single units and sensation: A neuron doctrine for perceptual psychology. *Perception*, 1: 371–394.

Barlow, H. B., 1995. The neuron in perception. In: Gazzaniga, M. S., ed., *The Cognitive Neurosciences*. MIT Press, Cambridge, MA.

Barnea, A., and Nottebohm, F., 1994. Seasonal recruitment of hippocampal neurons in adult free-ranging black-capped chickadees. *Proc. Natl. Acad. Sci. U.S.A.*, 91: 11217–11221.

Barnea, A., and Nottebohm, F., 1996. Recruitment and replacement of hippocampal neurons in young and adult chickadees: An addition to the theory of hippocampal learning. *Proc. Natl. Acad. Sci. U.S.A.*, 93: 714–718.

Bartlett, J., 1956. *Familiar Quotations*. Macmillan, London.

Bastholm, E., 1950. *The History of Muscle Physiology from the Natural Philosophers to Albrecht von Haller*. Munksgaard, Copenhagen.

Bax, D., 1979. *Hieronymus Bosch: His Picture Writing Deciphered*. N. A. Bax-Brotha, trans. A. A. Balkema, Lisse, Netherlands.

Bayer, S. A., and Altman, J., 2007. *Atlas of Human Central Nervous System*. 5 vols. CRC Press, Boca Raton, FL.

Baylis, G. C., and Driver, J., 2001. Shape-coding in IT cells generalizes over contrast and mirror reversal, but not figure-ground reversal. *Nat. Neurosci.*, 4: 937–942.

Bechterev, V. M., 1932. *General Principles of Human Reflexology.* International Press, New York.

Beevor, C. E., 1887. A further minute analysis by electric stimulation of the so-called motor region of the *cortex cerebri* in the monkey (*Macacus sinicus*). *Phil. Trans. R. Soc. Lond. B*, 179: 205–256.

Beevor, C. E., and Horsley, V., 1887. A minute analysis (experimental) of the various movements produced by stimulating in the monkey different regions of the cortical centre for the upper limb, as defined by Professor Ferrier. *Phil. Trans. R. Soc. Lond. B*, 178: 153–167.

Benton, A. L., 1975. Developmental dyslexia: Neurological aspects. In: Friedlander, W. J., ed., *Advances in Neurology,* vol. 7. Raven Press, New York.

Bernard, C., 1950. *Lettres Beaujolaises.* J. Godart, ed. Éditions du Cuvier, Villefranche-en-Beaujolais, France.

Bernard, C., 1961 [1865]. *Introduction to the Study of Experimental Medicine.* H. C. Greene, trans. Collier, New York.

Bernard, C., 1967. *Cahier Rouge.* H. H. Hoff, L. Guillemin, and R. Guillemin, trans. Schenkman, Cambridge, MA.

Bernard, C., 1974 [1878]. *Lectures on Phenomena Common to Animals and Plants.* H. E. Hoff, R. Guillemin, and L. Guillemin, trans. Charles C. Thomas, Springfield, IL.

Bernard, C., 1978 [1869–1878]. *Lettres Parisiennes: 1869–1878.* Jacqueline Sonolet and Fondation Marcel Merieux, Paris.

Bernier, P. J., Bedard, A., Vinet, J., Levesque, M. and Parent, A., 2002. Newly generated neurons in the amygdala and adjoining cortex of adult primates. *Proc. Natl. Acad. Sci. U.S.A.*, 99: 11464–11469.

Bhardwaj, R. D., Curtis, M. A., Spalding, K. L., Buchholz, B. A., Fink, D., Björk-Eriksson, T., Nordborg, C., Gage, F. H., Druid, H., Eriksson, P. S., and Frisén, J., 2006. Neocortical neurogenesis in humans is restricted to development. *Proc. Natl. Acad. Sci. U.S.A.*, 103: 12564–12568.

———

Biederman, I., and Cooper, E. E., 1991. Evidence for complete translation and reflectional invariance in visual object priming. *Perception*, 20: 585–593.

Blakemore, C., 1973a. The baffled brain. In: Gregory, R. L., and Gombrich, E. H., eds., *Illusion in Nature and Art.* Scribner's, New York.

Blakemore, C., 1973b. The language of vision. *New Sci.*, 58: 674–677.

Boas, F., 1955. *Primitive Art.* Dover, New York.

Boring, E. G., 1950. *A History of Experimental Psychology*, 2nd ed. Appleton-Century-Crofts, New York.

Bornstein, M. H., Gross, C. G., and Wolf, J. Z., 1978. Perceptual similarity of mirror images in infancy. *Cognition*, 6: 89–116.

Boss, B. D., Peterson, G. M., and Cowan, W. M., 1985. On the number of neurons in the dentate gyrus of the rat. *Brain Res.*, 338: 144–150.

Bowersock, G. W., 1969. *Greek Sophists in the Roman Empire.* Clarendon Press, Oxford.

Bradshaw, J. L., Nettleton, N. C., and Patterson, K., 1973. Identification of mirror-reversed and non-reversed facial profiles in same and opposite visual fields. *J. Exp. Psychol.*, 99: 42–8.

Brazier, M. A. B., 1959. The historical development of neurophysiology. In: Magoun, H. W., ed., *Handbook of Physiology*, section 1, vol. 1. American Physiological Society, Washington, DC.

Brazier, M. A. B., 1984. *A History of Neurophysiology in the 17th and 18th Centuries: From Concept to Experiment.* Raven Press, New York.

Brazier, M. A. B., 1988. *A History of Neurophysiology in the 19th Century.* Raven Press, New York.

Breasted, J. H., 1930. *The Edwin Smith Surgical Papyrus.* University of Chicago Press, Chicago.

Brecht, M., Schneider, M., Sakmann, B., and Margrie, T. W., 2004. Whisker movements evoked by stimulation of single pyramidal cells in rat motor cortex. *Nature*, 427: 704–710.

Broca, P., 1960 [1861]. Remarks on the seat of the faculty of articulate language, followed by an observation of aphemia. In: von Bonin, G., trans., *Some Papers on the Cerebral Cortex*. Charles C. Thomas, Springfield, IL.

Brooks, B., and Jung, R., 1973. Neuronal physiology of the visual cortex. In: Jung, R., ed., *Handbook of Sensory Physiology*, vol. VII/3B. Springer, Berlin.

Brown, T. G., and Sherrington, C. S., 1912. On the instability of a cortical point. *Proc. R. Soc. Lond. B*, 85: 250–277.

Brown, T. G., and Sherrington, C. S., 1915. Studies in the physiology of the nervous system. XXII. On the phenomenon of facilitation. 1. Its occurrence in reactions induced by stimulation of the "motor" cortex of the cerebrum in monkeys. *Q. J. Exp. Physiol.*, 9: 81–100.

Bruce, C., Desimone, R., and Gross, C. G., 1981. Visual properties of neurons in a polysensory area in superior temporal sulcus of the macaque. *J. Comp. Neurophysiol.*, 46: 369–384.

Bruel-Jungerman, E., Rampon, C., and Laroche, S., 2007. Adult hippocampal neurogenesis, synaptic plasticity and memory: Facts and hypotheses. *Rev. Neurosci*, 18: 93–114.

Bryans, W. A., 1959. Mitotic activity in the brain of the adult white rat. *Anat. Rec.*, 133: 65–71.

Burd, G. D., and Nottebohm, F., 1985. Ultrastructural characterization of synaptic terminals formed on newly generated neurons in a song control nucleus of the adult canary forebrain. *J. Comp. Neurol.*, 240: 143–152.

Burke, R. S., and Dallenbach, K. M., 1924. Position vs. intensity as a determinant of attention of left-handed observers. *Am. J. Psychol.*, 34: 267–269.

Burton, R., 1652. *The Anatomy of Melancholy*. Cripps, London.

Burton, R. F., trans., 1885. *The Book of the Thousand Nights and a Night*, vol. 5. Burton Club, London.

Burton, R. F., trans., 1886. *The Perfumed Garden of the Cheikh Nefzaoui*. Cosmopoli for the Kama Shasta Society, London.

Buswell, G. T., 1935. *How People Look at Pictures*. University of Chicago Press, Chicago.

Butler, A. B., and Hodos, W., 1996. *Comparative Vertebrate Neuroanatomy*. Wiley, New York.

Calder, R., 1970. *Leonardo and the Age of the Eye*. Heinemann, London.

Cameron, H. A., and Dayer, A. G., 2008. New interneurons in the adult neocortex: Small, sparse, but significant? *Biol. Psychiatry*, 63: 650–655.

Cameron, H. A., and McKay, R. D., 2001. Adult neurogenesis produces a large pool of new granule cells in the dentate gyrus. *J. Comp. Neurol.*, 435: 406–417.

Cameron, H. A., Wolley, C. S., McEwen, B. S., and Gould, E., 1993. Differentiation of newly born neurons and glia in the dentate gyrus of the adult rat. *Neuroscience*, 56: 337–344.

Cannon, W. B., 1963 [1929]. *Bodily Changes in Pain, Hunger, Fear and Rage*. Harper and Row, New York.

Cannon, W. B., 1963 [1932]. *The Wisdom of the Body*. Norton, New York.

Caramazza, A., 2000. The organization of conceptual knowledge in the brain. In: Gazzaniga, M. S., ed., *The New Cognitive Neurosciences*. MIT Press, Cambridge, MA.

Cazort, M., Kornell, M., and Roberts, K. B., 1996. *The Ingenious Machine of Nature*. National Gallery of Canada, Ottawa.

Cheney, P. D., and Fetz, E. E., 1985. Comparable patterns of muscle facilitation evoked by individual corticomotoneuronal (CM) cells and by single intracortical microstimuli in primates: Evidence for functional groups of CM cells. *J. Neurophysiol.*, 53: 786–804.

Churchland, P. S., 1986. *Neurophilosophy: Toward a Unified Science of the Mind-Brain*. MIT Press, Cambridge, MA.

Cinotti, M., 1969. *The Complete Works of Bosch*. Weidenfield and Nicholson, London.

Clark, K. M., 1978. *An Introduction to Rembrandt*. Murray, London.

Clarke, E., 1970. David Ferrier. In: Gillespie, C. C., ed., *The Dictionary of Scientific Biography*. Scribner's, New York.

Clarke, E., and Bearn, J. G., 1968. The brain "glands" of Malpighi elucidated by practical history. *J. Hist. Med. Allied Sci.*, 23: 309–330.

Clarke, E., and Dewhurst, K. E., 1972. *An Illustrated History of Brain Function*. University of California Press, Berkeley.

Clarke, E., and O'Malley, C. D., 1996. *The Human Brain and Spinal Cord: A Historical Study Illustrated by Writings from Antiquity to the Twentieth Century*. Norman, San Francisco.

Cline, R. H., 1972. Heart and eyes. *Romance Philol.*, 25: 263–297.

Cohen, I. B., 1985. *Revolution in Science*. Harvard University Press, Cambridge, MA.

Cohen, M. R., and Drabkin, I. E., 1958. *A Source Book in Greek Science*. Harvard University Press, Cambridge, MA.

Coleman, W., 1985. The cognitive basis of the discipline: Claude Bernard on physiology. *Isis*, 76: 49–70.

Collins, J., and Jin, D. Z., 2006. Grandmother cells and the storage capacity of the human brain. *Digital Library for Physics and Astronomy*, March. http://adsabs.harvard.edu.

Colombo, M., Colombo, A., and Gross, C. G., 2002. Bartolomeo Panizza's *Observations on the Optic Nerve* (1855). *Behav. Brain Res.*, 58: 529–539.

Colombo, M., and Gross, C. G., 1994. Responses of inferior temporal cortex and hippocampal neurons during delayed matching to sample in monkeys (*Macaca fascicularis*). *Behav. Neurosci.*, 108: 443–455.

Cooter, R. J., 1985. *The Cultural Meaning of Popular Science: Phrenology and the Organization of Consent in Nineteenth-Century Britain*. Cambridge University Press, Cambridge.

Coover, J. E., 1913. The feeling of being stared at. *Amer. J. Psychol.*, 24: 570–575.

Corballis, M. C., and Beale, I. L., 1976. *The Psychology of Left and Right*. Halstead, New York.

Corballis, M. C., Milner, A., and Morgan, M. J., 1971. The role of left-right orientation in inter-hemispheric matching of visual information. *Percept. Psychophys.*, 10: 385–388.

Coren, S., 1992. *The Left-Hander Syndrome: The Causes and Consequences of Left-Handedness*. Free Press, New York.

Corotto, F. S., Henegar, J. R., and Maruniak, J. A., 1994. Odor deprivation leads to reduced neurogenesis and reduced neuronal survival in the olfactory bulb of the adult mouse. *Neuroscience*, 61: 739–744.

Cottrell, J. E., and Winer, G. A., 1994. Development in the understanding of perception: The decline of extramission perception beliefs. *Dev. Psychol.*, 30: 218–228.

Cottrell, J. E., Winer, G. A., and Smith, M. C., 1996. Beliefs of children and adults about feeling stares of unseen others. *Dev. Psychol.*, 32: 50–61.

Couffer, J., 1992. *Bat Bomb: World War II's Other Secret Weapon*. University of Texas Press, Austin.

Cowey, A., 1994. Cortical visual areas and the neurobiology of higher visual processes. In: Farah, M. J., and Ratcliff, G., eds., *The Neuropsychology of High-Level Vision: Collected Tutorial Essays*. Erlbaum, Hillsdale, NJ.

Coxton, A., 1962. The Kissii art of trephining. *Guy's Hosp. Gaz.*, 76: 263–265.

Cranefield, P. F., 1974. *The Way In and the Way Out: François Magendie, Charles Bell and the Roots of Spinal Nerves*. Futura, Mount Kisco, NY.

Critchley, M., 1953. *The Parietal Lobes*. Arnold, London.

Critchley, M., 1970. *The Dyslexic Child*. Heinemann, London.

Curtis, M. A., Kam, M., Nannmark, U., Anderson, M. F., Axell, M. Z., Wikkelso, C., Holtås, S., van Roon-Mom, W. M., Björk-Eriksson, T., Nordborg, C., Frisén, J., Dragunow, M., Faull, R. L., and Eriksson, P. S., 2007. Human neuroblasts migrate to the olfactory bulb via a lateral ventricular extension. *Science*, 315: 1243–1249.

Dagi, T. F., 1997. The management of head trauma. In: Greenblatt, S. H., ed., *A History of Neurosurgery*. American Association of Neurological Surgeons, Park Ridge, IL.

Dallenbach, K. M., 1923. Position vs. intensity as a determinant of clearness. *Am. J. Psychol.*, 34: 282–286.

Darwin, C., 1871. *The Descent of Man and Selection in Relation to Sex*. Murray, London.

Dayer, A. G., Cleaver, K. M., Abouantoun, T., and Cameron, H. A., 2005. New GABAergic interneurons in the adult neocortex and striatum are generated from different precursors. *J. Cell. Biol.*, 168: 415–427.

Dean, A., 1946. *Fundamentals of Play Directing*. Rinehart, New York.

Delevoy, R. L., 1990. *Bosch*. Skira, Milan.

Deloulme, J. C., Lucas, M., Gaber, C., Bouillon, P., Keller, A., Eclancher, F., and Sensenbrenner, M., 1996. Expression of the neuron-specific enolase gene by rat oligoden-droglial cells during their differentiation. *J. Neurochem.*, 66: 936–945.

Descartes, R., 1972 [1662]. *Treatise on Man*. T. S. Hall, trans. Harvard University Press, Cambridge, MA.

Desimone, R., 1991. Face-selective cells in the temporal cortex of monkeys. *J. Cogn. Neurosci.*, 3: 1–8.

Desimone, R., 1996. Neural mechanisms for visual memory and their role in attention. *Proc. Natl. Acad. Sci. U.S.A.*, 93: 13494–13499.

Desimone, R., Albright, T. D., Gross, C. G., and Bruce, C., 1984. Stimulus-selective properties of inferior temporal neurons in the macaque. *J. Neurosci.*, 4: 2051–2062.

Dewhurst, K. E., 1982. Thomas Willis and the foundations of British neurology. In: Rose, F., and Bynum, W., eds., *Historical Aspects of the Neurosciences*. Raven Press, New York.

Diderot, D., and D'Alembert, J., 1751. *Encyclopedia ou Dictionnaire Raisonne des Sciences, des Artes, et des Metiers*. Pellet, Geneva.

Dimond, S. J., and Beaumont, J. G., eds., 1974. *Hemisphere Function in the Human Brain*. Elek Science, London.

Dimopoulous, V. G., Robinson, J. S. III, and Fountas, K. N., 2008. The pearls and pitfalls of skull trephination as described in the Hippocratic treatise "On Head Wounds." *J. Hist. Neurosci.*, 17: 131–140.

Dobson, J. F., 1926–1927. Erasistratus. *Proc. R. Soc. Med.*, 20: 825–832.

Dobson, R., 2000. Doctors warn of the dangers of trepanning. *Br. Med. J.*, 320: 602 (on-line only). http://www.bmj.com/cgi/content/full/320/7235/602/c.

Donaldson-Evans, L. K., 1980. *Love's fatal glance: A study of eye imagery in the poets of the École Lyonnaise*. Romance Monographs, University of Missouri, Columbia, MS.

Dudok van Heel, S. A. C., 1998. *Nicolaes Tulp: The Life and Work of an Amsterdam Physician and Magistrate in the 17th Century*. K. Gribling, trans. Six Art Promotion, Amsterdam.

Dundes, A., ed., 1981. *The Evil Eye: A Folklore Case Book*. Garland, New York.

Dupré, S., ed., 2005. *Early Science and Medicine, Special Issue: Optics, Instruments and Painting 1420–1720. Reflections on the Hockney-Falco Thesis*. 10: 125–339.

Eckenhoff, M. F., and Rakic, P., 1988. Nature and fate of proliferative cells in the hippocampal dentate gyrus during the life span of the rhesus monkey. *J. Neurosci.*, 8: 2729–2747.

Elliott, P., 1987. Vivisection and the emergence of experimental physiology in nineteenth-century France. In: Rupke, N. A., ed., *Vivisection in Historical Perspective*. Croom Helm, Beckenham, UK.

Ellis, H. D., and Shepherd, J., 1975. Recognition of upright and inverted faces presented in the left and right visual fields. *Cortex*, 11: 3–7.

Elston, M. A., 1987. Women and anti-vivisection in Victorian England, 1870–1900. In: Rupke, N. A., ed., *Vivisection in Historical Perspective*. Croom Helm, New York.

Eriksson, P. S., Perfilieva, E., Björk-Eriksson, T., Alborn, A., Nordborg, C., Peterson, D. A., and Gage, F. H., 1998. Neurogenesis in the adult human hippocampus. *Nat. Med.*, 4: 1313–1317.

Falco, C. M., 2007a. Computer vision and art. *IEEE Multimedia*, 14: 8–11.

Falco, C. M., 2007b. Ibn al-Haytham and the origins of modern image analysis. Presented at the International Conference on Information Sciences, Signal Processing and Its Applications, Sharjah, U.A.E.

Farrington, B., 1932. The last chapter of the *De Fabrica* of Vesalius entitled: Some observations on the dissection of living animals. *Trans. R. Soc. S. Afr.*, 20: 1–14.

Fearing, F., 1970. *Reflex Action: A Study in the History of Physiological Psychology*. Williams and Wilkins, Baltimore.

Fernando, H. R., and Finger, S., 2003. Ephraim George Squier's Peruvian skull and the discovery of cranial trepanation. In: Arnott, R., Finger, S., and Smith, C. U. M., eds., *Trepanation: History, Discovery, Theory*. Swets and Zeitlinger, Lisse, Netherlands.

Ferrier, D., 1873. Experimental researches in cerebral physiology and pathology. *West Riding Lunatic Asylum Medical Reports*, 3: 30–96.

Ferrier, D., 1874a. Pathological illustrations of brain function. *West Riding Lunatic Asylum Medical Reports*, 4: 30–62.

Ferrier, D., 1874b. The localization of function in the brain [abstract]. *Proc. R. Soc. Lond.*, 22: 228–232.

Ferrier, D., 1874c. The localization of function in the brain. Manuscript, Archives of the Royal Society, A. P. 56.2 228–232.

Ferrier, D., 1874–1875. Experiments on the brain of monkeys: No. 1. *Proc. R. Soc. Lond.*, 23: 409–430.

Ferrier, D., 1875. The Croonian Lecture: Experiments on the brain of monkeys (second series). *Phil. Trans. R. Soc. Lond.*, 165: 433–488.

Ferrier, D., 1876. *Functions of the Brain*. Putnam, New York.

Ferrier, D., 1878. *The Localisation of the Cerebral Disease*. Smith Elder, London.

Ferrier, D., 1886. *Functions of the Brain*, 2nd ed. Smith Elder, London.

Ferrier, D., 1890. Cerebral localization in its practical relations. *Brain*, 12: 36–58.

Finger, S., 1994. *Origins of Neuroscience: A History of Explorations into Brain Function*. Oxford University Press, New York.

Finger, S., 2000. *Minds Behind the Brain: The Pioneers and Their Discoveries*. Oxford University Press, New York.

Finger, S., and Clower, W. T., 2003. On the birth of trepanation: The thoughts of Paul Broca and Victor Horsley. In: Arnott, R., Finger, S., and Smith, C. U. M., eds., *Trepanation: History, Discovery, Theory*. Swets and Zeitlinger, Lisse, Netherlands.

Fiser, J., and Biederman, I., 2001. Invariance of long-term visual primitives to scale, reflection, translation, and hemisphere. *Vision Res.*, 41: 221–234.

Flamm, E. S., 1997. From signs to symptoms: The neurosurgical management of head trauma from 1717 to 1867. In: Greenblatt, S. H., ed., *A History of Neurosurgery*. American Association of Neurological Surgeons, Park Ridge, IL.

Flint, A. J. Jr., 1878. Claude Bernard and his physiological works. *Am. J. Med. Sci.*, 76: 161–173.

Fonberg, E., 1974. Professor Jerzy Konorski. *Acta Neurobiol. Exp.*, 34: 655–664.

Foster, M., 1899. *Claude Bernard.* Unwin, London.

Fredericq, L., 1973. The influence of the environment on the composition of blood of aquatic animals. In: Langley, L. L., ed., *Homeostasis: Origins of the Concept.* Dowden, Hutchinson and Ross, Stroudsburg, PA.

Freeman, K., 1954. *The Pre-Socratic Philosophers.* Blackwell, Oxford.

French, R. D., 1975. *Antivivisection and Medical Science in Victorian Society.* Princeton University Press, Princeton.

Frisby, J. P., 1980. *Seeing.* Oxford University Press, Oxford.

Frisch, K. von, 1950. *Bees, Their Vision, Chemical Senses and Language.* Cornell University Press, Ithaca, NY.

Fritsch, G. T., 1912. *Das Haupthaar und seine Bildungstatte bei den Rassen des Menschen.* 2 vols. G. Reimer, Berlin.

Fritsch, G. T., and Hitzig, E., 1960 [1870]. On the electrical excitability of the cerebrum. In: von Bonin, G., trans., *Some Papers on the Cerebral Cortex.* Charles C. Thomas, Springfield, IL.

Fry, C. C., 1946. Art in the history of medicine: The sixteenth century cures for lunacy. *Amer. J. Psychiatry*, 103: 351–353.

Fulton, J. F., 1949a. *Functional Localization in the Frontal Lobes and Cerebellum.* Clarendon Press, Oxford.

Fulton, J. F., 1949b. *Physiology of the Nervous System*, 3rd ed. Oxford University Press, New York.

Gaffron, M., 1950. Right and left in pictures. *Art Q.*, 13: 312–331.

Galambos, R., 1942. The avoidance of obstacles by flying bats: Spallanzani's ideas (1794) and later theories. *ISIS*, 34: 132–140.

Galambos, R., 1995. Robert Galambos. In: Squire, L., ed., *The History of Neuroscience in Autobiography*, vol. 1. Academic Press, New York.

Galen, 1962 [2nd C.]. *On Anatomical Procedures: The Later Books*. W. L. H. Duckworth, trans. Cambridge University Press, Cambridge.

Galen, 1968 [2nd C.]. *On the Usefulness of the Parts of the Body*. M. May, trans. Cornell University Press, Ithaca, NY.

Galen, 1978–1984 [2nd C.]. *On the Doctrines of Hippocrates and Plato*. P. De Lacy, trans. Akademie-Verlag, Berlin.

Galen, 1979 [2nd C.]. *On Prognosis. English and Greek*. V. Nutton, trans. Akademie-Verlag, Berlin.

Galen, 1988 [2nd C.]. *On Examinations by which the Best Physicians Are Recognized*. A. Z. Iskandar, trans. Akademie-Verlag, Berlin.

Gall, F. J., and Spurzheim, J. C., 1810–1819. *Anatomie et physiologie du système nerveux en général, et du cerveau en particulier, avec des observations sur la possibilité de reconnaître plusieurs dispositions intellectuelles et morales de l'homme et des animaux, par la configuration de leurs têtes.* 4 vols. with atlas of 100 engraved plates. (Gall is sole author of vols. 3 and 4.) Schoell, Paris.

Gall, F. J., and Spurzheim, J. C., 1835. *On the Function of the Brain and Each of Its Parts: With Observations on the Possibility of Determining the Instincts, Propensities and Talents, or the Moral and Intellectual Disposition of Men and Animals, by the Configuration of the Brain and Head.* W. Lewis Jr., trans. Marsh, Capen and Lyon, Boston.

Gardner, H., ed., 1957. *The Metaphysical Poets*. Penguin, Harmondsworth, UK.

Gardner, H., 1975. *Gardner's Art through the Ages*, 6th ed., revised by H. de la Croix and R. G. Tansey. Harcourt Brace Jovanovich, New York.

Gardner, H., 1985. *The Mind's New Science: A History of the Cognitive Revolution*. Basic Books, New York.

Gardner, M., 1969. *The Ambidextrous Universe: Left, Right and the Fall of Parity*. Mentor, New York.

Gazzaniga, M. S., Ivry, R. B., and Mangun, G. R., 1998. *Cognitive Neuroscience: The Biology of the Mind*. Norton, New York.

Georgopoulos, A. P., Ashe, J., Smyrnis, N., and Taira, M., 1992. The motor cortex and the coding of force. *Science*, 256: 1692–1695.

Georgopoulos, A. P., Schwartz, A. B., and Kettner, R. E., 1986. Neuronal population coding of movement direction. *Science*, 233: 1416–1419.

Gibson, W. W., 1973. *Hieronymus Bosch*. Thames and Hudson, London.

Gifford, E. S. Jr., 1958. *The Evil Eye: Studies in the Folklore of Vision*. Macmillan, New York.

Gilbert, C., and Bakan, P., 1973. Visual asymmetry in perception of faces. *Neuropsychology*, 11: 355–362.

Glickstein, M., 1985. Ferrier's mistake. *Trends Neurosci.*, 8: 341–344.

Goldman, S. A., and Nottebohm, F., 1983. Neuronal production, migration and differentiation in a vocal control nucleus of the adult female canary brain. *Proc. Natl. Acad. Sci. U.S.A.*, 80: 2390–2394.

Goodrich, J. T., 1997. Neurosurgery in the ancient and medieval worlds. In: Greenblatt, S. H., ed., *A History of Neurosurgery*. American Association of Neurological Surgeons, Park Ridge, IL.

Gordon, I. E., and Gardner, D., 1974. Responses to altered pictures. *Br. J. Psychol.*, 65: 243–251.

Goss, C. M., 1966. On anatomy of nerves by Galen of Pergamon. *Am. J. Anat.*, 118: 327–336.

Gould, E., 2006. Non-synaptic plasticity in the hippocampus. In: Anders, P., Morris, R., Amaral, D., Bliss, T., and O'Keefe, T., eds., *The Hippocampus Book*. Oxford University Press, Oxford.

Gould, E., 2007. How widespread is adult neurogenesis in mammals? *Nat. Rev. Neurosci.*, 8: 481–488.

Gould, E., Reeves, A. J., Fallah, M., Tanapat, P., Gross, C. G., and Fuchs, E., 1999a. Hippocampal neurogenesis in Old World primates. *Proc. Natl. Acad. Sci. U.S.A.*, 96: 5263–5267.

Gould E., Reeves, A. J., Graziano, M. S., and Gross, C. G., 1999b. Neurogenesis in the neocortex of adult primates. *Science*, 286: 548–552.

Gould E., Tanapat, P., McEwen, B. S., Flugge, G., and Fuchs, E., 1998. Proliferation of granule cell precursors in the dentate gyrus of adult monkeys is diminished by stress. *Proc. Natl. Acad. Sci. U.S.A.*, 95: 3168–3171.

Gould E., Vail, N., Wagers, M., and Gross, C. G., 2001. Adult-generated hippocampal and neocortical neurons in macaques have a transient existence. *Proc. Natl. Acad. Sci. U.S.A.*, 98: 10910–10917.

Grabman, J. M., 1975. The witch of Mallegem. *J. Hist. Med. Allied Sci.*, 30: 384–385.

Grande, F., 1967. Introduction to the symposium. In: Grande, F., and Visscher, M. B., eds., *Claude Bernard and Experimental Medicine*. Schenkman, Cambridge, MA.

Graziano, M. S. A., 2006. The organization of behavioral repertoire in motor cortex. *Annu. Rev. Neurosci.*, 29: 105–134.

Graziano, M. S. A., and Aflalo, T. N., 2007. Rethinking cortical organization: Moving away from discrete areas arranged in hierarchies. *Neuroscientist*, 13: 138–147.

Graziano, M. S. A., Taylor, C. S. R., and Moore, T., 2002a. Complex movements evoked by microstimulation of precentral cortex. *Neuron*, 34: 841–851.

Graziano, M. S. A., Taylor, C. S. R., Moore, T., and Cooke, D. F., 2002b. The cortical control of movement revisited. *Neuron,* 36: 349–362.

Gregg, V. R., Winer, G. A., Cottrell, J. E., Hedman, K. E., and Fournier, J. S., 2001. The persistence of a misconception about vision after educational interventions. *Psychon. Bull. Rev.*, 8: 622–626.

Griffin, D. R., 1934. Marking bats. *J. Mammal.*, 15: 202–207.

Griffin, D. R., 1958. *Listening in the Dark: The Acoustic Orientation of Bats and Men*. Yale University Press, New Haven.

Griffin, D. R., 1964. *Bird Navigation*. Doubleday, Garden City, NJ.

Griffin, D. R., 1976. *The Question of Animal Awareness*. Rockefeller University Press, New York.

Griffin, D. R., 1980. Early history of research on echolocation. In: Busnel, R.-G., and Fish, J. F., eds., *Animal Sonar Systems*. Plenum, New York.

Griffin, D. R., 1984. *Animal Thinking*. Harvard University Press, Cambridge, MA.

Griffin, D. R., 1985. Reflections of an experimental naturalist. In: Dewsbury, D. A., ed., *Leaders in the Study of Animal Behavior*. Bucknell University Press, Lewisburg, PA.

Griffin, D. R., 1992. *Animal Minds*. University of Chicago Press, Chicago.

Griffin, D. R., 1998. Donald R. Griffin. In: Squire, L., ed., *The History of Neuroscience in Autobiography, vol. 2*. Academic Press, NY.

Grmek, M. D., 1970a. François Magendie. In: Gillespie, C. C., ed., *Dictionary of Scientific Biography*. Scribner's, New York.

Grmek, M. D., 1970b. Claude Bernard. In: Gillespie, C. C., ed., *Dictionary of Scientific Biography*. Scribner's, New York.

Gross, C. G., 1968. Review of J. Konorski, *Integrative Activity of the Brain* (1967). *Science*, 160: 652–653.

Gross, C. G., 1978. Inferior temporal lesions do not impair discrimination of rotated patterns in monkeys. *J. Comp. Physiol. Psychol.*, 92: 1095–1109.

Gross, C. G., 1981. Ibn-al-Haytham on eye and brain, vision and perception. *Bull. Islam. Med.*, 1: 309–312.

Gross, C. G., 1992. Representation of visual stimuli in inferior temporal cortex. *Phil. Trans. R. Soc. Lond. B*, 335: 3–10.

Gross, C. G., 1993a. The Hippocampus minor and man's place in nature: A case study in the social construction of neuroanatomy. *Hippocampus*, 3: 403–415.

Gross, C. G., 1993b. Huxley vs. Owen: The hippocampus minor and evolution. *Trends Neurosci.*, 16: 493–498.

Gross, C. G., 1994. How inferior temporal cortex became a visual area. *Cereb. Cortex*, 5: 455–469.

Gross, C. G., 1995. Aristotle on the brain. *Neuroscientist*, 1: 245–250.

Gross, C. G., 1997a. Emanuel Swedenborg: A neuroscientist before his time. *Neuroscientist*, 3: 142–147.

Gross, C. G., 1997b. Leonardo da Vinci on the brain and eye. *Neuroscientist*, 3: 347–354.

Gross, C. G., 1998a. *Brain, Vision, Memory: Tales in the History of Neuroscience*. MIT Press, Cambridge, MA.

Gross, C. G., 1998b. Galen and the squealing pig. *Neuroscientist*, 4: 216–221.

Gross, C. G., 1998c. Rembrandt's "The Anatomy Lesson of Dr. Deijman." *Trends Neurosci.*, 1998, 21: 237–240.

Gross, C. G., 1998d. Claude Bernard and the constancy of the internal environment. *Neuroscientist*, 4: 380–385.

Gross, C. G., 1999a. The fire that comes from the eye. *Neuroscientist*, 5: 58–64.

Gross, C. G., 1999b. A hole in the head. *Neuroscientist*, 5: 263–269.

Gross, C. G., 1999c. "Psychosurgery" in Renaissance art. *Trends Neurosci.*, 22: 429–431.

Gross, C. G., 1999d. Phrenology. In: Adelman, G., ed., *Encyclopedia of Neuroscience*, 2nd ed. Birkhauser, Boston.

Gross, C. G., 2000a. Neurogenesis in the adult brain: Death of a dogma. *Nat. Rev. Neurosci.*, 1: 67–73.

Gross, C. G., 2000b. Coding for visual categories in the human brain. *Nat. Neurosci.*, 3: 855–856.

Gross, C. G., 2002. The Genealogy of the "Grandmother Cell." *Neuroscientist*, 8: 512–518.

Gross, C. G., 2003. Trepanation from the Paleolithic to the Internet. In: Arnott, R., Finger, S., and Smith, C. U. M., eds., *Trepanation: History, Discovery, Theory*. Swets and Zeitlinger, Lisse, Netherlands.

Gross, C. G., 2005. Donald R. Griffin 1915–2003. *Biographical Memoirs of the National Academy of Sciences* 86: 1–20.

Gross, C. G., 2007. The discovery of motor cortex and its background in the 18th and early 19th centuries. *J. Hist. Neurosci.*, 16: 320–331.

Gross, C. G., 2009. Three before their time: Neuroscientists whose ideas were ignored by their contemporaries. *Exp. Brain Res.*, 192: 321–334.

Gross, C. G., Bender, D. B., and Rocha-Miranda, C. E., 1969. Visual receptive fields of neurons in inferotemporal cortex of the monkey. *Science*, 166: 1303–1306.

Gross, C. G., and Bornstein, M. H., 1978. Left and right in science and art. *Leonardo*, 11: 29–38.

Gross, C. G., and Mishkin, M., 1977. The neural basis of stimulus equivalence across retinal translation. In: Harnad, S., Doty, R., Jaynes, J., Goldstein, L., and Krauthammer, G., eds., *Lateralization in the Nervous System*. Academic Press, New York.

Gross, C. G., Rocha-Miranda, C. E., and Bender, D. B., 1972. Visual properties of neurons in inferotemporal cortex of the macaque. *J. Neurophysiol.*, 35: 96–111.

Grounds, J. G., 1958. Trephining of the skull amongst the Kissii. *East Afr. Med. J.*, 35: 369–373.

Grundfest, H., 1963. The different careers of Gustav Fritsch. *J. Hist. Med. Allied Sci.* 18: 125–129.

Grusser, O-J., and Hagner, M., 1990. On the history of deformation phosphenes and the idea of internal light generated in the eye for the purpose of vision. *Doc. Ophthalmol.*, 74: 57–85.

Grusser, O-J., Grusser-Cornehls, U., Hagner, M., and Przybyszewski, A., 1989. Purkyne's description of pressure phosphenes and modern neurophysiological studies on the generation of phosphenes by eyeball deformation. *Physiol. Bohemoslov.*, 38: 289–309.

Guanzhong, L., 1991. *Three Kingdoms*. M. Roberts, trans. University of California, Berkeley.

Gur, R. E., 1975. Conjugate lateral eye movements as an index of hemispheric activation. *J. Pers. Soc. Psychol.*, 31: 751–757.

Guthrie, D., 1945. *History of Medicine*. Nelson, London.

Hahnloser, R. H., Koshevnikov, A. A., and Fee, M. S., 2002. An ultra-sparse code underlies the generation of neural sequences in a songbird. *Nature*, 419: 65–70.

Haldane, J. S., 1931. *The Philosophical Basis of Biology*. Hodder and Stoughton, London.

Hamilton, A., 1901. The division of differentiated cells in the central nervous system of the white rat. *J. Comp. Neurol.*, 11: 297–320.

Hamilton, C. R., and Tieman, S. B., 1973. Interocular transfer of mirror image discriminations by chiasm-sectioned monkeys. *Brain Res.*, 64: 241–255.

Hampson, J., 1987. Legislation: A practical solution to the vivisection dilemma? In: Rupke, N. A., ed., *Vivisection in Historical Perspective*. Croom Helm, New York.

Handyside, J., trans., 1929. *Kant's Inaugural Dissertation and Early Writings on Space*. Open Court, Chicago.

Hansen, J. V., 1996. Resurrecting death: Anatomical art in the cabinet of Dr. Frederick Ruysch. *Art Bull.*, 78: 663–679.

Harcum, E. R., and Finkel, M. E., 1963. Explanation of Mishkin and Forgay's result as a directional reading conflict. *Can. J. Psychol.*, 17: 224–233.

Harnad, S., Doty, R., Jaynes, J., Goldstein, L., and Krauthammer, G., eds., 1977. *Lateralization in the Nervous System*. Academic Press, New York.

Harris, L., 1995. *The Secret Heresy of Hieronymus Bosch*. Floris Books, Edinburgh.

Hasselmo, M. E., Rolls, E. T., and Baylis, G. C., 1989. The role of expression and identity in the face-selective responses of neurons in the temporal and visual cortex of the monkey. *Behav. Brain Res.*, 32: 203–218.

Haxby, J. V., Grady, C. L., Ungerleider, L. G., and Horwitz, B., 1991. Mapping the functional neuroanatomy of the intact human brain with brain work imaging. *Neuropsychologia*, 29: 539–555.

Hecaen, H., and Sauquet, J., 1971. Cerebral dominance in left-handed subjects. *Cortex*, 7: 19–48.

Hecksher, W. S., 1958. *Rembrandt's Anatomy of Dr. Tulp*. New York University Press, New York.

Held, J. S., 1991. *Rembrandt Studies*. Princeton University Press, Princeton.

Henderson, L. J., 1928. *Blood: A Study of General Physiology*. Yale University Press, New Haven.

Henderson, L. J., 1935. *Pareto's General Sociology: A Physiologist's Interpretation*. Harvard University Press, Cambridge, MA.

Henderson, L. J., 1958 [1913]. *The Fitness of the Environment*. Beacon, Boston.

Henderson, L. J., 1961 [1927]. Introduction. In: Bernard, C., *Introduction to the Study of Experimental Medicine*. Collier, New York.

Herrick, C. L., 1882. Neurologists and neurological laboratories. 1. Professor Gustav Fritsch. *J. Comp. Neurol.*, 2: 84–88.

Hess, E., 1965. Attitude and pupil size. *Sci. Am.*, 212: 46–54.

Hippocrates, 1927 [4th C. bce]. On wounds in the head. In: Jones, W. H. S., and Withington, E. T., trans., *Hippocrates*. Heinemann, London.

Hippocrates, 1950 [4th C. bce]. On the sacred disease. In: Chadwick, J., ed., *The Medical Works of Hippocrates*. Blackwell, Oxford.

His, W., 1904. *Die Entwicklung des menschlichen Gehirns*. Hirzel, Leipzig.

Hitzig, E., 1870. Über die galvanischen Schwindelempfindungen und eine neue Methode galvanischer Reizung der Augenmuskeln. *Klin. Wochenschr.*, 7: 137–138.

Hitzig, E., 1874. *Untersuchungnen über das Gehirn: Abhandlungen physiologischen und pathologischen Inhalts*. Hirschwald, Berlin.

Hitzig, E., 1900. Hughlings Jackson and the cortical motor centres in the light of physiological research. *Brain*, 23: 545–581.

Hockney, D., 2001. *Secret Knowledge: Rediscovering the Lost Techniques of the Old Masters*. Viking Studio, New York.

Holmes, E. J., and Gross, C. G., 1984. Effects of inferior temporal lesions on discrimination of stimuli differing in orientation. *J. Neurosci.*, 4: 3063–3068.

Holmes, F. L., 1963. Claude Bernard and the milieu interieur. *Arch. Int. Hist. Sci.*, 16: 369–376.

Holmes, F. L., 1965. Contributions of marine biology to the development of the concept of the milieu interieur. *Vie Milieu*, 19 (Suppl.): 321–335.

Holmes, F. L., 1967. Origins of the concept of milieu interieur. In: Grande, F., and Visscher, M. B., eds., *Claude Bernard and Experimental Medicine*. Schenkman, Cambridge, MA.

Holmes, F. L., 1974. *Claude Bernard and Animal Chemistry*. Harvard University Press, Cambridge, MA.

Honig, L. S., Herrmann, K., and Shatz, C. J., 1996. Developmental changes revealed by immunohistochemical markers in human cerebral cortex. *Cereb. Cortex*, 6: 794–806.

Horsley, V., and Schäfer, E. A., 1883. Experimental researches in cerebral physiology. *Proc. R. Soc. Lond.*, 36: 437–442.

Horsley, V., and Schäfer, E. A., 1888. A record of experiments upon the functions of the cerebral cortex. *Phil. Trans. R. Soc. Lond. B*, 179: 1–45.

Hua-Tao, 1993 [2nd C.]. *Master Hua's Classic of the Central Viscera*. Yang Shou-zhong, trans. Blue Poppy Press, Boulder, CO.

Hubel, D. H., and Wiesel, T. N., 1962. Receptive fields, binocular interaction and functional architecture in the cat's visual cortex. *J. Physiol. (Lond.)*, 160: 106–154.

Hubel, D. H., and Wiesel, T. N., 1965. Receptive fields and functional architecture in two nonstriate visual areas (18 and 19) of the cat. *J. Neurophysiol.*, 28: 229–289.

Huxley, T. H., 1863. *Evidence as to Man's Place in Nature*. Macmillan, London.

IJpma, F. F., van de Graaf, R. C., Nicolai, J. P., and Meek, M. F., 2006. The anatomy lesson of Dr. Nicolaes Tulp by Rembrandt (1632): A comparison of the painting with dissected left forearm of a Dutch male cadaver. *J. Hand Surg. [Am].*, 31: 882–891.

Isaac, E., 1980. The Princeton collection of Ethiopic manuscripts. *Princeton University Library Chronicle*, 42: 32–52.

Jackowe, D. J., Moore, M. K., Bruner, A. E., and Fredieu, J. R., 2007. New insight into the enigmatic white cord in Rembrandt's *The Anatomy Lesson of Dr. Nicolaes Tulp*. *J. Hand Surg.*, 32: 1471–1476.

Jackson, J. H., 1870. A Study of Convulsions. *Transactions of the St. Andrews Medical Graduates' Association*, vol. 3. Reprinted in: Jackson, J. H. (1875), *Selected Writings*. Vol. 1, *On Epilepsy and Epileptiform Convulsions*. J. Taylor, ed. Basic Books, New York.

Jackson, J. H., 1875. On the anatomical and physiological localization of movements in the brain. Reprinted in: Jackson, J. H. (1875), *Selected Writings*. Vol. 1, *On Epilepsy and Epileptiform Convulsions*. J. Taylor, ed. Basic Books, New York.

Jackson, J. H., 1958. *Selected Writings*. Vol. 1, *On Epilepsy and Epileptiform Convulsions*. J. Taylor, ed. Basic Books, New York.

Jacobson, M., 1970. *Developmental Neurobiology*. Holt, Rinehart and Winston, New York.

James, W., 1890. *Principles of Psychology*. MacMillan, London.

Jaynes, C. J., 1976. *The Origin of Consciousness in the Breakdown of the Bicameral Mind*. Houghton Mifflin, Boston.

Jefferson, Sir G. J., 1960. *Selected Papers*. Pitman Medical Publishing, New York.

Jensen, B. T., 1952a. Left-right orientation in profile drawing. *Am. J. Psychol.*, 65: 80–83.

Jensen, B. T., 1952b. Reading habits and left-right orientation in profile drawing by Japanese children. *Am. J. Psychol.*, 65: 306–307.

Jung, M. W., and McNaughton, B. L., 1993. Spatial selectivity of unit activity in the hippocampal granular layer. *Hippocampus*, 3: 165–182.

Kakei, S., Hoffman, D., and Strick, P., 1999. Muscle and movement representations in the primary motor cortex. *Science*, 285: 2136–2139.

Kanwisher, N., McDermott, J., and Chun, M. M., 1997. The fusiform face area: A module in human extrastriate cortex specialized for face perception. *J. Neurosci.*, 17: 4302–4311.

Kaplan, M. S., 1981. Neurogenesis in the 3-month-old rat visual cortex. *J. Comp. Neurol.*, 195: 323–338.

Kaplan, M. S., 1983. Proliferation of subependymal cells in the adult primate CNS: Differential uptake of DNA labeled precursors. *J. Hirnforsch.*, 23: 23–33.

Kaplan, M. S., 1984. Mitotic neuroblasts in the 9-day-old and 11-month-old rodent hippocampus. *J. Neurosci.*, 4: 1429–1441.

Kaplan, M. S., 1985. Formation and turnover of neurons in young and senescent animals: An electron microscopic and morphometric analysis. *Ann. N. Y. Acad. Sci.*, 457: 173–192.

Kaplan M. S., 2001. Environment complexity stimulates visual cortex neurogenesis: Death of a dogma and a research career. *Trends Neurosci.*, 24: 617–620.

Kaplan, M. S., and Hinds, J. W., 1977. Neurogenesis in the adult rat: Electron microscopic analysis of light radioautographs. *Science*, 197: 1092–1094.

Keele, K. D., 1957. *Anatomies of Pain.* Blackwell, Oxford.

Kemp, M., 1990. *The Science of Art: Optical Themes in Western Art from Brunelleschi to Seurat.* Yale University Press, New Haven.

Kempermann, G., Kuhn, H. G., and Gage, F. H., 1997. More hippocampal neurons in adult mice living in an enriched environment. *Nature*, 386: 493–495.

Kershman, J., 1938. The medulloblast and the medulloblastoma. *Arch. Neurol. Psychiatry,* 40: 937–967.

Kinsbourne, M., 1974. Direction of gaze and distribution of cerebral thought processes. *Neuropsychology*, 12: 279–281.

Kinsbourne, M., ed., 1976. *Hemispheric Asymmetries of Function.* Cambridge University Press, Cambridge.

Kirn, J. R., and Nottebohm, F., 1993. Direct evidence for loss and replacement of projection neurons in adult canary brain. *J. Neurosci.*, 13: 1654–1663.

Klein, E., Burdon-Sanderson, J., Foster, M., and Brunton, T. L., 1873. *Handbook for the Physiological Laboratory.* J. and A. Churchill, London.

Klein, E., Langley, J. N., and Schäfer, E. A., 1883. On the cortical areas removed from the brain of a dog, and from the brain of a monkey. *J. Physiol.*, 4: 231–247.

Klein, H. A., ed., 1963. *Graphic Worlds of Peter Bruegel the Elder.* Dover, New York.

Koelliker, A., 1896. *Handbuch der Gewebelehre des Menschen.* Engelmann, Leipzig.

Konorski, J., 1948. *Conditioned Reflexes and Neuron Organization.* Cambridge University Press, Cambridge.

Konorski, J., 1967. *Integrative Activity of the Brain: An Interdisciplinary Approach.* University of Chicago Press, Chicago.

Konorski, J., 1974. Jerzy Konorski. In: Lindzey, G., ed., *A History of Psychology in Autobiography.* Prentice Hall, Englewood Cliffs, NJ.

Kornack, D. R., and Rakic, P., 1999. Continuation of neurogenesis in the hippocampus of the adult macaque monkey. *Proc. Natl. Acad. Sci. U.S.A.,* 96: 5768–5773.

Kosinski, K. D., 1999. *The Painted Bird.* Bantam, New York.

Kreiman, G., Koch, C., and Fried, I., 2000. Category-specific visual responses of single neurons in the human medial temporal lobe. *Nat. Neurosci.,* 3: 946–953.

Kreiman, G., Koch, C., and Fried, I., 2001. Single neuron responses in humans during binocular rivalry and flash suppression. *Abstr. Soc. Neurosci.,* 27.

Kruger, L., 1963. François Pourfour du Petit, 1664–1741. *Exp. Neurol.,* 7: 2–5.

Kruger, L., 2005. The scientific impact of Dr. N. Tulp, portrayed in Rembrandt's "Anatomy Lesson." *J. Hist. Neurosci.,* 14: 85–92.

Kuhn, T. S., 1970. *The Structure of Scientific Revolutions.* 2nd ed. University of Chicago Press, Chicago.

Kuhn, T. S., Dickinson-Anson, H., and Gage, F. H., 1996. Neurogenesis in the dentate gyrus of the adult rat: Age-related decrease of neuronal progenitor proliferation. *J. Neurosci.,* 16: 2027–2033.

Kuntz, E., 1953. Eduard Hitzig. In: Haymaker, W., ed., *The Founders of Neurology.* Charles C. Thomas, Springfield, IL.

Landauer, M. S., 1969. The orientation of forms in abstract art. *Proc. A. P. A.,* 4: 475–476.

Langley, L. L., 1973. Introduction and comments. In: Langley, L. L., ed., *Homeostasis: Origins of the Concept*. Dowden, Hutchinson and Ross, Stroudsburg, PA.

Lee, M. K., Tuttle, J. B., Rebhun, L. I., Cleveland, D. W., and Frankfurter, A., 1990. The expression and posttranslational modification of a neuron-specific beta-tubulin isotype during chick embryogenesis. *Cell Motil. Cytoskeleton*, 17: 118–132.

Lehman, R. A., and Spencer, D. D., 1973. Mirror image shape discrimination: Interocular reversal of responses in the optic chiasm sectioned monkey. *Brain Res.*, 52: 233–241.

Leonard, C. M., Rolls, E. T., and Baylis, G. C., 1985. Neurons in the amygdala of the monkey with responses selective for faces. *Behav. Brain Res.*, 15: 159–176.

Leonardo da Vinci, 1970 [14th C.]. *Notebooks*. J. P. Richter, ed. Dover, New York.

Lettvin, J. Y., Maturana, H. R., McCulloch, W. S., and Pitts, W. H., 1959. What the frog's eye tells the frog's brain. *Proc. Inst. Radio Eng.*, 47: 1940–1951.

Lettvin, J. Y., Maturana, H. R., McCulloch, W. S., and Pitts, W. H., 1961. Two remarks on the visual system of the frog. In: Rosenblith, W. A., ed., *Symposium on Principles of Sensory Communication*. MIT Press, Cambridge, MA.

Leuner, B., Gould, E., and Shors, T. J., 2006. Is there a link between adult neurogenesis and learning? *Hippocampus*, 16: 216–224.

Levi, G., 1898. Sulla cariocinesi delle cellule nervose. *Riv. Patol. Nerv. Ment.*, 3: 97–113.

Levy, J., 1976. Lateral dominance and aesthetic preference. *Neuropsychology*, 14: 431–435.

Lewes, G. H., 1877. *The Physical Basis of Mind*. Trubner, London.

Leyton, A. S. F., and Sherrington, C. S., 1916. Observations on the excitable cortex of the chimpanzee, orang-utan, and gorilla. *Q. J. Exp. Physiol.*, 11: 135–222.

Lind, L. R., ed., 1954. *Lyric Poetry of the Italian Renaissance*. Yale University Press, New Haven.

Lindberg, D. C., 1976. *Theories of Vision from Al-Kindi to Kepler*. Chicago, University of Chicago Press.

Lindberg, D. C., 1992. *The Beginnings of Western Science*. Chicago, University of Chicago Press.

Lindeboom, G. A., 1977. Medical aspects of Rembrandt's Anatomy Lesson of Dr. Tulp. *Janus*, 64: 179–203.

Lisowski, F. P., 1967. Prehistoric and early historic trepanation. In: Brothwell, D., and Sandison, A. T., eds., *Diseases in Antiquity*. Charles C. Thomas, Springfield, IL.

Lloyd, G. E. R., 1975. Alcmaeon and the early history of dissection. *Sudhoffs Arch.*, 59: 113–147.

Lloyd, G. E. R., 1978. The Hippocratic question. *Class. Q.*, 28: 202–222.

Loeb, J., 1900. *Comparative Physiology of the Brain and Comparative Psychology*. Putnam, New York.

Logothetis, N. K., Guggenberger, H., Peled, S., and Pauls, J., 1999. Functional imaging of the monkey brain. *Nat. Neurosci.*, 2: 555–562.

Logothetis, N. K., Pauls, J., and Poggio, T., 1995. Shape representation in the inferior temporal cortex of monkeys. *Curr. Biol.*, 5: 552–563.

Logothetis, N. K., and Sheinberg, D. L., 1996. Visual object recognition. *Annu. Rev. Neurosci.*, 19: 577–621.

Lois, C. Alvarez-Buylla, A., 1994. Long-distance neuronal migration in the adult mammalian brain. *Science*, 264: 1145–1148.

Longrigg, J., 1993. *Greek Rational Medicine: Philosophy and Medicine from Alcmaeon to the Alexandrians*. Routledge, London.

Lovell, M. S., 1998. *A Rage to Live: A Biography of Richard and Isabel Burton*. Norton, New York.

Lu, G., and Needham, J., 1980. *Celestial Lancets: A History and Rationale of Acupuncture and Moxa*. Cambridge University Press, Cambridge.

Luciani, L., and Tamburini, A., 1879. Ricerche sperimentali sulle funzioni del cervello, II: Centri psico-sensori corticali. *Riv. Sper. Freniatr. Med. Leg.*, 5: 1–76.

Mach, E., 1914. *The Analysis of Sensations and the Relation of the Physical to the Psychical*. S. Waterlow, ed. Open Court, Chicago.

Macallum, A. B., 1926. The paleochemistry of the body fluids and tissues. *Physiol. Rev.*, 6: 316–357.

Magavi, S. S., Leavitt, B. R., and Macklis, J. D., 2000. Induction of neurogenesis in the neocortex of adult mice. *Nature*, 405: 951–955.

Magner, L. N., 1992. *A History of Medicine*. Dekker, New York.

Majno, G., 1975. *The Healing Hand: Man and Wound in the Ancient World*. Harvard University Press, Cambridge, MA.

Makita, K., 1968. The rarity of reading disability in Japanese children. *J. Orthopsychiatry*, 38: 599–614.

Malpighi, M., 1996 [1666]. *De cerebri cortice*. Montius, Bologna. Quotation translated in: Clarke, E., and O'Malley C. D., eds., *The Human Brain and Spinal Cord: A Historical Study Illustrated by Writings from Antiquity to the Twentieth Century*. Norman, San Francisco.

Mandairon, N., Sacquet, J., Garcia, R., Ravel, N., Jourdan, F., and Didier, A., 2006. Neurogenic correlates of an olfactory discrimination task in the adult olfactory bulb. *Eur. J. Neurosci.*, 24: 3578–3588.

Manni, E., and Petrosini, L., 1994. Contributions by Bartolomeo Panizza to the anatomy and physiology of some cranial nerves. *J. Hist. Neurosci.*, 3: 187–197.

Manuel, D., 1987. Marshall Hall (1790–1857): Vivisection and the development of experimental physiology. In: Rupke, N. A., ed., *Vivisection in Historical Perspective*. Croom Helm, Beckenham, UK.

Margetts, E. L., 1967. Trepanation of the skull by the medicine-men of primitive cultures, with particular reference to present day native East African practice. In: *Diseases in Antiquity*. D. Brothwell and A. T. Sandison, eds. Charles C. Thomas, Springfield, IL.

Margetts, E. L., 1998. Trepanning the skull in East Africa, a follow-up of Gusii (Kisii) folk surgery after 35 years. *Anthologica Medica Santoriniana*, 1: 5–10.

Marr, D., 1982. *Vision: A Computational Investigation into the Human Representation and processing of visual information*. W. H. Freeman, San Francisco.

Martin, A., Ungerleider, L. G., and Haxby, J. V., 2000. Category specificity and the brain: The sensory/motor model of semantic representations of objects. In: Gazzaniga, M. S., ed., *The New Cognitive Neurosciences*. MIT Press, Cambridge, MA.

Mazzarello, P., and Della Sala, S., 1993. The demonstration of the visual area by means of the atrophic degeneration method in the work of Bartolomeo Panizza (1855). *J. Hist. Neurosci.*, 2: 315–322.

McCloskey, M., and Kargon, R., 1988. The meaning and use of historical models in the study of intuitive physics. In: Strauss, S., ed., *Ontogeny, Phylogeny, and Historical Development*. Ablex, Norwood, NJ.

McIlwain, J. T., 2006. Brain and mind in Anglo-Saxon medicine. *Viator*, 37: 103–112.

McManus, I. C., 2002. *Right Hand, Left Hand: The Origins of Asymmetry in Brains, Bodies, Atoms, and Cultures*. Harvard University Press, Cambridge, MA.

McManus, I. C., and Humphrey, N. K., 1973. Turning the left cheek. *Nature*, 243: 271–272.

Mellick, S. A., 2007. Dr. Nicolaes Tulp of Amsterdam, 1593–1674: Anatomist and doctor of medicine. *ANZ J. Surg.*, 77: 1102–1109.

Menden, J. M., 1969. Operation for stones in the head. *J. Hist. Med. Allied Sci.*, 24: 210–211.

Menezes, J. R., and Luskin, M. B., 1994. Expression of neuron-specific tubulin defines a novel population in the proliferative layers of the developing telencephalon. *J. Neurosci.*, 14: 5399–5416.

Mettler, F. A., and Mettler, C. C., 1945. Historic development of knowledge relating to cranial trauma. In: Browda, J., Rabiner, A. M., and Mettler, F. A., eds., *Trauma of the Central Nervous System*. Williams and Wilkins, Baltimore.

Meyer, A. M., 1971. *Historical Aspects of Cerebral Anatomy*. Oxford University Press, London.

Middelkoop, N., 1994. *De anatomische le van Dr. Deijman*. Amsterdams Historisch Museum, Amsterdam.

Miller, E. K., Li, L., and Desimone, R., 1991. A neural mechanism for working and recognition memory in inferior temporal cortex. *Science*, 254: 1377–1379.

Mills, A. R., 1989. Rembrandt's painting of the Anatomy Lesson of Dr. Tulp. *Proc. R. Coll. Physicians Edinb.*, 19: 327–330.

Mintum, J. E., Teschwind, D. H., Fryer, H. J. L., and Hockfield, S., 1995. Early postmitotic neurons transiently express TOAD-64, a neural specific protein. *J. Comp. Neurol.*, 355: 369–379.

Mishkin, M., 1966. Visual mechanisms beyond the striate cortex. In: Russell, R. W., ed., *Frontiers in Physiological Psychology*. Academic Press, New York.

Mitchell, J. F., 1999. The people with holes in their heads. In: Mitchell, F., ed., *Eccentric Lives and Peculiar Notions*. Harcourt Brace Jovanovich, San Diego.

Miyashita, Y., 1988. Neuronal correlate of visual associative long-term memory in the primate temporal cortex. *Nature*, 335: 817–820.

Moeller, S., Freiwald, W. A., and Tsao, D. Y., 2008. Patches with links: A unified system for processing faces in the macaque temporal lobe. *Science*, 320: 1355–1359.

Money, J., ed., 1962. *Reading Disability*. Johns Hopkins, Baltimore.

Morgan, C. L., 1895. *Introduction to Comparative Psychology*. Scribner's, New York.

Morris, M., ed. and trans., 1951. *Philosophical Writings of Leibnitz*. J. M. Dent, London.

Mullen, R. J., Buck, C. R., and Smith, A. M., 1992. NeuN, a neuronal specific nuclear protein in vertebrates. *Development*, 116: 201–211.

Müller, J. M., 1965 (1838). On the specific energies of nerves. In: Herrnstein, R., and Boring, E., ed., *A Sourcebook of the History of Psychology*. Harvard University Press, Cambridge, MA.

Munk, H., 1878. Weitere Mittheilungen zur Physiologie der Grosshirnrinde. *Verh. Physiol. Ges. Berl.*, 581–594.

Munk, H., 1960 [1881]. Uber die Funktionen der Grosshirnrinde. In: von Bonin, trans., *Some Papers on the Cerebral Cortex*. Charles C. Thomas, Springfield, IL.

Nemesius, 1955 [4th C. BCE]. On the nature of man. In: Telfer, W., ed., *Cyril of Jerusalem and Nemesius of Emesa*. Westminster Press, Philadelphia.

Neuberger, M., 1981. *The Historical Development of Experimental Brain and Spinal Cord Physiology Before Flourens*. E. Clarke, trans. Johns Hopkins University Press, Baltimore.

New York Academy of Medicine, 1865. Stated meeting, December 6. *Bull. N. Y. Acad. Med.*, 1862–1866; 2: 530–547.

Noton, D., and Stark, L., 1971. Scanpaths in saccadic eye movements while viewing and recognizing patterns. *Vision Res.*, 11: 929–942.

Nottebohm, F., 1985. Neuronal replacement in adulthood. *Ann. N. Y. Acad. Sci.*, 457: 143–161.

Nottebohm, F., 1989. Hormonal regulation of synapses and cell number in the adult canary brain and its relevance to theories of long-term memory storage. In: Lakoski, J. M., Perez-Polo, J. R., and Rossin, D. K., eds., *Neural Control of Reproductive Function*. A. R. Liss, New York.

Nottebohm, F., 1996. The King Solomon Lectures in Neuroethology. A white canary on Mount Acropolis. *J. Comp. Physiol. A*, 179: 149–156.

Nowakowski, R. S., Lewin, S. B., and Miller, M. W., 1989. Bromodeoxyuridine immunohistochemical determination of the lengths of the cell cycle and the DNA-synthetic phase for an anatomically defined population. *J. Neurocytol.*, 18: 311–318.

Nutton, V., 1984. Galen in the eyes of his contemporaries. *Bull. Hist. Med.*, 58: 315–324.

Nutton, V., 1995. Galen *ad multos annos*. *Dynamis*, 15: 25–40.

Nutton, V., ed., 2002. *The Unknown Galen*. Institute of Classical Studies, London.

O'Brian, Patrick, 1984. *The Far Side of the World*. Norton, New York.

Ojemann, J. G., Ojemann, G. A., and Lettich, E., 1992. Neuronal activity related to faces and matching in human right nondominant temporal cortex. *Brain*, 115: 1–13.

Olmsted, E. H., 1967. Historical phases in the influence of Bernard's scientific generalizations in England and America. In: Grande, F., and Visscher, M. B., eds., *Claude Bernard and Experimental Medicine*. Schenkman, Cambridge, MA.

Olmsted, J. M. D., 1939. *Claude Bernard, Physiologist*. Cassell, London.

Olmsted, J. M. D., 1944. *François Magendie: Pioneer in Experimental Physiology and Scientific Medicine in XIX Century France*. Schuman, New York.

Olmsted, J. M. D., and Olmsted, E. H., 1952. *Claude Bernard and the Experimental Method in Medicine*. Schuman, New York.

Oppe, A. P., 1944. Right and left in Raphael's cartoons. *J. Warburg and Courtauld Inst.*, 7: 82–94.

Orton, S. T., 1937. *Reading, Writing, and Speech Problems in Children*. Norton, New York.

Oxford Dictionary of Quotations, 1959. Oxford University Press, Oxford.

Ozer, M. N., 1966. The British vivisection controversy. *Bull. Hist. Med.*, 40: 158–167.

Panizza, B., 1855. Osservazioni sul nervo ottico. *Gior. I. R. Ist. Lomb. Sci. Lett. Arti.*, 7: 237–252.

Panizza, B., 1856. Osservazioni sul nervo ottico. *Mem. I. R. Ist. Lomb. Sci. Lett. Arti.*, 5: 375–390.

Paton, J. A., and Nottebohm, F. N., 1984. Neurons generated in the adult brain are recruited into functional circuits. *Science*, 225: 1046–1048.

Pavlov, I. P., 1929. *Lectures on Conditioned Reflexes*. W. H. Gantt, trans. International Publishers, New York.

Penfield, W., and Rasmussen, T., 1950. *The Cerebral Cortex of Man: A Clinical Study of Localisation of Function*. Macmillan, New York.

Perera, T. D., Park, S., and Nemirovskaya, Y., 2008. Cognitive role of neurogenesis in depression and antidepressant treatment. *Neuroscientist*, 14: 326–338.

Perrett, D. I., Rolls, E. T., and Caan, W., 1982. Visual neurons responsive to faces in the monkey temporal cortex. *Exp. Brain Res.*, 47: 329–342.

Perrett, D. I., Smith, P. A., Potter, D. D., Mistlin, A. J., Head, A. S., and Milner, A. D., 1985. Visual cells in the temporal cortex sensitive to face view and gaze direction. *Proc. R. Soc. Lond. B*, 223: 293–317.

Petit, A., 1987. Claude Bernard and the history of science. *Isis*, 78: 201–219.

———

Phillips, C. G., 1975. Laying the ghost of "muscles versus movements." *Can. J. Neurol. Sci.*, 2: 209–218.

Piaget, J., 1979. *The Child's Conception of the World.* J. Tomlinson and A. Tomlinson, trans. Littlefield, Adams, Totowa, NJ.

Pinsk, M. A., DeSimone, K., Moore, T., Gross, C. G., and Kastner, S., 2005. Representations of faces and body parts in macaque temporal cortex: A functional MRI study. *Proc. Natl. Acad. Sci. U.S.A.*, 102: 6996–7001.

Plato, 1920 [4th C. BCE]. *The Dialogues of Plato.* B. Jowett, trans. Random House, New York.

Plato, 1959 [4th C. BCE]. *Timaeus.* F. M. Cornford, trans. New York, Bobbs-Merrill.

Polyak, S. L., 1957. *The Vertebrate Visual System.* University of Chicago Press, Chicago.

Prelog, V., 1976. Chirality in chemistry. *Science*, 193: 17–24.

Pribram, K. H., and Mishkin, M., 1955. Simultaneous and successive visual discrimination by monkeys with inferotemporal lesions. *J. Comp. Physiol. Psychol.*, 48: 198–202.

Puce, A., Allison, T., Gore, J. C., and McCarthy, G., 1995. Face-sensitive regions in human extrastriate cortex studied by functional MRI. *J. Neurophysiol.*, 74: 1192–1199.

Quiroga, R. Q., Kreiman, G., Koch, C., and Fried, I., 2007. Sparse but not "grandmother-cell" coding in the medial temporal lobe. *Trends Cogn. Sci.*, 12: 87–91.

Quiroga, R. Q., Reddy, L., Kreiman, G., Koch, C., and Fried I., 2005. Invariant visual representation by single neurons in the human brain. *Nature*, 435: 1102–1107.

Rakic, P., 1985a. Limits of neurogenesis in primates. *Science*, 227: 1054–1056.

Rakic, P., 1985b. DNA synthesis and cell division in the adult primate brain. *Ann. N. Y. Acad. Sci.*, 457: 193–211.

Ramón y Cajal, S., 1928 [1913]. *Degeneration and Regeneration of the Nervous System.* R. M. Day, trans. Oxford University Press, London.

Ramón y Cajal, S., 1999 [1904]. *Texture of the Nervous System of Man and the Vertebrates.* P. Pasik and T. Pasik, trans. Springer, Vienna.

Ranson, S. W., 1920. *The Anatomy of the Nervous System.* Saunders, Philadelphia.

Rashed, R., 2002. Portraits of science. A polymath in the 10th century. *Science,* 297: 773.

Reconstructing Rembrandt: The grim reaper exposed, 2000. *IC Reporter,* July 3.

Reed, T., 2000. Painting is an illusion. *Times Higher Education Supplement,* December 15.

Remnant, P., 1963. Incongruent counterparts and absolute space. *Mind,* 72: 393–399.

Richards, S., 1987. Vicarious suffering, necessary pain: Physiological method in late nineteenth-century Britain. In: Rupke, N. A., ed., *Vivisection in Historical Perspective.* New York, Croom Helm.

Richter, C. R., 1927. Animal behavior and internal drives. *Q. Rev. Biol.,* 2: 307–343.

Robin, E. D., ed., 1979. *Claude Bernard and the Internal Environment: A Memorial Symposium.* M. Dekker, New York.

Rocca, J., 2003a. Galen and the uses of trepanation. In: Arnott, R., Finger, S., and Smith, C. U. M., eds., *Trepanation: History, Discovery, Theory.* Swets and Zeitlinger, Lisse, Netherlands.

Rocca, J., 2003b. *Galen on the Brain: Anatomical Knowledge and Physiological Speculation in the Second Century A.D.* Brill, Leiden, Netherlands.

Rodman, H. R., O'Scalaidhe, S. P., and Gross, C. G., 1993. Response properties of neurons in temporal cortical visual areas of infant monkeys. *J. Neurophysiol.,* 70: 1115–1136.

Rollenhagen, J. E., and Olson, C. R., 2000. Mirror-image confusion in single neurons of the macaque inferotemporal cortex. *Science,* 287: 1506–1508.

Rolls, E. T., 1984. Neurons in the cortex of the temporal lobe and in the amygdala of the monkey with responses selective for faces. *Human Neurobiology,* 3: 209–222.

Rolls, E. T., and Tovee, M. J., 1995. Sparseness of the neuronal representation of stimuli in the primate temporal visual cortex. *J. Neurophysiol.,* 73: 713–726.

Romanes, G. J., 1882. *Animal Intelligence.* Kegan, Paul, Trench, London.

Rosen, G., 1939. Trepanation in Cornish miners. *Ciba Symposium,* 1: 197.

Rosenberg, J., 1968. *Rembrandt: Life and Work*. Phaidon, Vienna.

Rosenblith, W. A., ed., 1961. *Symposium on Principles of Sensory Communication*. MIT Press, Cambridge, MA.

Rosenblueth, A., Wiener, N., and Bigelow, J., 1943. Behavior, teleology and purpose. *Philos. Sci.*, 10: 18–24.

Rosenzweig, M. R., Leiman, A. L., and Breedlove, S. M., 1999. *Biological Psychology: An Introduction to Behavioral, Cognitive, and Clinical Neuroscience*. 2nd ed. Sinauer Associates, Sunderland, MA.

Ross, B. M., 1966. Minimal familiarity and left-right judgments of paintings. *Percept. Mot. Skills*, 22: 105–106.

Rosser, A. E., Tyers, P., ter Borg, M., Dunnett, S. B., and Svendsen, C. N., 1997. Co-expression of MAP-2 and GFAP in cells developing from rat EGF responsive precursor cells. *Brain Res. Dev. Brain Res.*, 98: 291–295.

Roth, P., 1969. *Portnoy's Complaint*. Random House, New York.

Roy, N. S., Wang, S., Jiang, L., Kang, J., Benraiss, A., Harrison-Restelli, C., Fraser, R. A. R., Couldwell, W. T., Kawaguchi, A., Okano, H., Nedergaard, M., and Goldman, S. A., 2000. *In vitro* neurogenesis by progenitor cells isolated from the adult human hippocampus. *Nat. Med.*, 6: 271–277.

Rudel, R. G., and Teuber, H.-L., 1963. Discrimination of direction of line in children. *J. Comp. Physiol. Psychol.*, 56: 892–898.

Ruisinger, M. M., 2003. Lorenz Heister (1683–1758) and the "Bachmann case": Social setting and medical practice of trepanation in eighteenth-Century Germany. In: Arnott, R., Finger, S., and Smith, C. U. M., eds., *Trepanation: History, Discovery, Theory*. Swets and Zeitlinger, Lisse, Netherlands.

Runyan, C. A., Weickert, C. S., and Saunders, R. C., 2006. Adult neurogenesis and immediate early gene response to working memory stimulation in the primate prefrontal cortex. *Soc. Neurosci. Abstr.*, 318.10.

Rupke, N., 1987. Pro-vivisection in England in the early 1880s: Arguments and motives. In: Rupke, N. A., ed., *Vivisection in Historical Perspective*. Croom Helm, New York.

Rupp, J. C. C., 1992. Michel Foucault, body politics and the rise and expansion of modern anatomy. *J. Hist. Soc.*, 5: 31–60.

Sabra, A. I., 1983. Abu 'Ali al-Hasan bin al Hasan bin al-haythem (Alhazen). In: Hayes, J. R., ed., *The Genius of Arab Civilization: Source of Renaissance*. MIT Press, Cambridge, MA.

Sarton, G., 1954. *Galen of Pergamon*. University of Kansas Press, Lawrence.

Sarton, G., 1959. *A History of Science*. Vol. 1, *Ancient Science through the Golden Age of Greece*. Harvard University Press, Cambridge, MA.

Saul, F. P., and Saul, J. M. 1997. Trepanation: Old world and new world. In: Greenblatt, S. H., ed., *A History of Neurosurgery*. American Association of Neurological Surgeons, Park Ridge, IL.

Saunders, J. B. deC. M., and O'Malley, C. D., 1950. *The Illustrations from the Works of Andreas Vesalius of Brussels*. World Publishing, Cleveland.

Scalaidhe, S. P., Wilson, F. A., and Godman-Rakic, P. S., 1999. Face-selective neurons during passive viewing and working memory performance of rhesus monkeys: Evidence for intrinsic specialization of neuronal coding. *Cereb. Cortex*, 9: 459–475.

Schama, S., 1987. *The Embarrassment of Riches: An Interpretation of Dutch Culture in the Golden Age*. Knopf, New York.

Schaper, A., 1897. Die frühesten differenzierungsvorgange im centralnervensystem. *Arch. f. Entw.-Mech. Organ.*, 5: 81–132.

Schiller, F., 1965. The rise of the "enteroid processes" in the 19th century: Some landmarks in cerebral nomenclature. *Bull. Hist. Med.*, 39: 326–338.

Schiller, J., 1967. Claude Bernard and vivisection. *J. Hist. Med. Allied Sci.*, 22: 246–260.

Schiller, F., 1992. *Paul Broca, Founder of French Anthropology, Explorer of the Brain*. Oxford University Press, Oxford.

Schlomoh d'Arles, G. B., 1953 [13th C.]. *The Gate of Heaven*. F. S. Bodenheimer, trans. Kiryath Sepher, Jerusalem.

———

Schmechel, D. E., Brightman, M. W., and Marangos, P. J., 1980. Neurons switch from non-neuronal enolase to neuron-specific enolase during differentiation. *Brain Res.*, 190: 195–214.

Schmitt, F. O., and Worden, F. G., eds., 1974. *The Neurosciences: Third Study Program.* MIT Press, Cambridge, MA

Scholzen, T., and Gerdes, J., 2000. The Ki-67 protein: From the known and the unknown. *J. Cell Physiol.*, 182: 311–322.

Schupbach, W., 1978. A new look at The Cure of Folly. *Med Hist.*, 22: 267–281.

Schwartz, E. L., Desimone, R., Albright, T. D., and Gross, C. G., 1983. Shape recognition and inferior temporal neurons. *Proc. Natl. Acad. Sci. U.S.A.*, 80: 5776–5778.

Scott, S. H., and Kalaska, J. F., 1997. Reaching movements with similar hand paths but different arm orientations. I. Activity of individual cells in motor cortex. *J. Neurophysiol.*, 77: 826–852.

Scultetus, J., 1655. *Armamentarium Chirurgicum.* Kühnen, Ulm.

Seki, T., and Arai, Y., 1995. Age-related production of new granule cells in the adult dentate gyrus. *Neuroreport*, 6: 2479–2482.

Seki, T., and Arai, Y., 1999. Temporal and spatial relationships between PSA-NCAM-expressing, newly generated granule cells, and radial glia-like cells in the adult dentate gyrus. *J. Comp. Neurol.*, 410: 503–513.

Sensenbrenner, M., Lucas, M., and Deloulme, J. C., 1997. Expression of two neuronal markers, growth-associated protein 43 and neuron-specific enolase, in rat glial cells. *J. Mol. Med.*, 75: 653–663.

Sergent, J., and Signoret, J. L., 1992. Functional and anatomical decomposition of face processing: Evidence from prosopagnosia and PET study of normal subjects. *Phil. Trans. R. Soc. Lond. B*, 335: 55–61.

Shadlen, M. N., Britten, K. H., Newsome, W. T., and Movshon, J. A., 1996. A computational analysis of the relationship between neuronal and behavioral responses to visual motion. *J. Neurosci.*, 16: 1486–1510.

Shankweiler, D. P., 1963. A study of developmental dyslexia. *Neuropsychology*, 1: 267–286.

Shapiro, M., 1970. On some problems in the semiotics of visual art: Field and vehicle in image-signs. In: Greimas, A. J., and Jacobson, R., eds., *Sign, Language, Culture*. Mouton, The Hague.

Sherrington, C. S., 1940. *Man on His Nature*. The University Press, Cambridge.

Sherrington, C. S., 1961 [1906]. *The Integrative Action of the Nervous System*. Yale University Press, New Haven.

Shors, T. J., 2008. From stem cells to grandmother cells: How neurogenesis relates to learning and memory. *Cell Stem Cell*, 3: 253–258.

Sidman, R. L., Miale, I. L., and Feder, N., 1959. Cell proliferation and migration in the primitive ependymal zone: An autoradiographic study of histogenesis in the nervous system. *Exp. Neurol.*, 1: 322–333.

Siegel, R. E., 1968. *Galen's System of Physiology and Medicine*. S. Karger, New York.

Siegel, R. E., 1973. *Galen on Psychology, Psychopathology, and Function and Diseases of the Nervous System*. S. Karger, New York.

Sigerist, H. E., 1933. *The Great Doctors*. E. C. Paul, trans. Norton, New York.

Sigerist, H. E., 1951. *A History of Medicine*. Vol. 1, *Primitive and Archaic Medicine*. Oxford University Press, Oxford.

Sigerist, H. E., 1961. *A History of Medicine*. Vol. 2, *Early Greek, Hindu, and Persian Medicine*. Oxford University Press, Oxford.

Simpson, D., 2007. Nicolaes Tulp and the golden age of the Dutch Republic. *ANZ J. Surg.*, 77: 1095–1101.

Singer, C., 1957. *A Short History of Anatomy and Physiology from the Greeks to Harvey*. Dover, New York.

Skinner, B. F., 1953. *Science and Human Behavior*. Macmillan, New York.

Smart, I., 1961. The subependymal layer of the mouse brain and its cell production as shown by autography after [H3]-thymidine injection. *J. Comp. Neurol.*, 116: 325–347.

Smith, A. M., 2005. Reflections on the Hockney-Falco thesis: Optical theory and artistic practices in the fifteenth and sixteenth centuries. In: Dupré, S., ed., *Early Science and Medicine, Special Issue: Optics, Instruments and Painting 1420–1720, Reflections on the Hockney-Falco Thesis*, 10: 163–185.

Smith, E. S., 1971. Galen's account of the cranial nerves and the autonomic nervous system. *Clio Med.*, 6: 77–89, 173–194.

Smith, W. D., 1979. *The Hippocratic Tradition*. Cornell University Press, Ithaca, NY.

Snyder, J., ed., 1973. *Bosch in Perspective*. Prentice Hall, Upper Saddle River, NJ.

So, K., Moriya, T., Nishitani, S., Takahashi, H., and Shinohara, K., 2008. The olfactory conditioning in the early postnatal period stimulated neural stem-progenitor cells in the subventricular zone and increased neurogenesis in the olfactory bulb of rats. *Neuroscience*, 151: 120–128.

Spallanzani, L., 1932 [1794]. *Opere di Lazzaro Spallanzani*. Vol. 3. Milan, Ulrico Hoepli.

Spence, S., 1996. "Lo cop mortal": The evil eye and the origins of courtly love. *Romanic Rev.*, 87: 307–319.

Spillane, J. D., 1981. *The Doctrine of the Nerves: Chapters in the History of Neurology*. Oxford University Press, Oxford.

Spurzheim, J. G. 1834. *Phrenology; or The Doctrine of the Mental Phenomenon*. 3rd ed. Marsh, Capen, and Lyon, Boston.

Squier, E. G., 1877. *Peru: Incidents of Travel and Exploration in the Land of the Incas*. Henry Holt, New York.

Stanfield, B. B., and Trice, J. E., 1988. Evidence that granule cells generated in the dentate gyrus of adult rats extend axonal projections. *Exp. Brain Res.*, 72: 399–406.

Starling, E. H., 1909. *The Fluids of the Body*. Keener, Chicago.

Storandt, M., 1974. Recognition across visual fields with mirror-image stimuli. *Percept. Mot. Skills*, 39: 762.

Stork, D. G., and Duarte, M. F., 2007. Computer vision, image analysis, and Master Art: Part 3. *IEEE Multimedia*, 14: 14–18.

Strick, P. L., 2002. Stimulating research on motor cortex. *Nat. Neurosci.*, 5: 714–715.

Suga, N., and Ma, X., 2003. Multiparametric corticofugal modulation and plasticity in the auditory system. *Nat. Rev. Neurosci.*, 4: 783–794.

Sugita, N., 1918. Comparative studies on the growth of the cerebral cortex. *J. Comp. Neurol.*, 29: 61–117.

Sutherland, N. S., 1960. Visual discrimination of orientation by octopus: Mirror images. *Br. J. Psychol.*, 51: 9–18.

Swartz, P., and Hewitt, D., 1970. Lateral organization in pictures and aesthetic preference. *Percept. Mot. Skills*, 30: 991–1007.

Swindler, M. H., 1929. *Ancient Painting.* Yale University Press, New Haven.

Symmes, J. A., and Rapoport, J. L., 1972. Unexpected reading failure. *Amer. J. Orthopsychiatry*, 42: 82–91.

Tamburini, A., 1880. Rivendicazione al Panizza della scoperta del centro visivo corticale. *Riv. Sper. Freniatr. Med. Leg.* 6: 153–154.

Tanaka, K., 1996. Inferotemporal cortex and object vision. *Annu. Rev. Neurosci.*, 19: 109–139.

Taylor, C. S. R., and Gross, C. G., 2003. Twitches versus Movements: A story of motor cortex. *Neuroscientist*, 9: 332–342.

Tee, K. S., and Riesen, A. H., 1974. Visual right-left confusions in animals and man. In: Newton, G., and Riesen, A. H., eds., *Advances in Psychobiology,* vol 2. Wiley, New York.

Temkin, O., 1946a. The philosophical background of Magendie's physiology. *Bull. Hist. Med.*, 20: 10–13.

Temkin, O., 1946b. Materialism in French and German physiology of the early nineteenth century. *Bull. Hist. Med.*, 20: 322–327.

Temkin, O., 1971. *The Falling Sickness: A History of Epilepsy from the Greeks to the Beginnings of Modern Neurology.* 2nd ed. Johns Hopkins Press, Baltimore.

Temkin, O., 1973. *Galenism: Rise and Decline of a Medical Philosophy.* Cornell University Press, Ithaca, NY.

Theophrastus, 1917 [4th BCE]. On the senses. In: Stratton, G. M., trans., *Theophrastus and the Greek physiological psychology Before Aristotle*. Allen and Unwin, London.

Thompson, C. J. S., 1938. The evolution and development of surgical instruments. IV. The trepan. *Br. J. Surg.*, 25: 726–734.

Thompson, D. W., 1969. *Growth and Form*. Cambridge University Press, Cambridge.

Thompson, L. T., and Best, P. J., 1989. Place cells and silent cells in the hippocampus of freely-behaving rats. *J. Neurosci.*, 9: 2382–90.

Tieleman, T., 1996. *Galen and Chrysippus on the Soul*. Brill, Leiden, Netherlands.

Tieleman, T., 2002. Galen on the seat of the intellect: Anatomical experiments and philosophical tradition. In: Tuplin, C. J., and Rihill, T. E., eds., *Science and Mathematics in Ancient Greek Culture*. Oxford University Press, Oxford.

Tinbergen, N., 1951. *The Study of Instinct*. Oxford University Press, Oxford.

Titchener, E. B., 1898. The "feeling of being stared at." *Science*, 8: 895–897.

Tsao, D. Y., Freiwald, W. A., Knutsen, T. A., Mandeville, J. B., and Tootell, R. B., 2003. Faces and objects in macaque cerebral cortex. *Nat. Neurosci.*, 6: 989–995.

Tsao, D. Y., Freiwald, W. A., Tootell, R. B., and Livingstone, M. S., 2006. A cortical region consisting entirely of face-selective cells. *Science*, 311: 670–674.

Tsao, D. Y., and Livingstone, M. S., 2008. Mechanisms of face perception. *Annu. Rev. Neurosci.*, 31: 411–437.

Tsao, D. Y., Moeller, S., and Freiwald, W. A.,2008. Comparing face patch systems in macaques and humans. *Proc. Natl. Acad. Sci. U.S.A.*, 105: 19514–19519.

Tyson, E., 1699. *Orang-outang, sive,* Homo sylvestris, *or, The Anatomy of a Pygmie Compared with that of a Monkey, an Ape, and a Man: To Which is Added, a Philological Essay concerning the Pygmies, the Cynocephali, the Satyrs and Sphinges of the Ancients: Wherein It Will Appear that They Are All either Apes or Monkeys, and Not Men, as Formerly Pretended*. Bennett and Brown, London.

Valenstein, E. S., 1997. History of Psychosurgery. In: Greenblatt, S. H., ed., *A History of Neurosurgery*. American Association of Neurological Surgeons, Park Ridge, IL.

———

Van Zoest, W., Giesbrecht, B., Enns, J. T., and Kingstone, A., 2006. New reflections on visual search: Interitem symmetry matters! *Psychol. Sci.*, 17: 535–542.

Viets, H. R., 1938. West Riding, 1871–1876. *Bull. Hist. Med.*, 6: 477–487.

Virtanen, R., 1960. *Claude Bernard and His Place in the History of Ideas.* University of Nebraska Press, Lincoln.

Von Gudden, J. B. A., 1870. Experimentaluntersuchungen bei das peripherischer und zentrale Nervensystem. *Arch. Psychiatr. Nervenkr.*, 2: 693–723.

Von Staden, H., 1989. *Herophilus: The Art of Medicine in Early Alexandria.* Cambridge University Press, Cambridge.

Walker, A. E., 1938. *The Primate Thalamus.* University of Chicago Press, Chicago.

Walker, A. E., 1998. *The Genesis of Neuroscience.* American Association of Neurological Surgeons, Park Ridge, IL.

Walsh, J., 1926. Galen's discovery and promulgation of the function of the recurrent laryngeal nerve. *Ann. Med. Hist.*, 8: 176–184.

Walsh, J., 1934–1939. Galen's writings and influences inspiring them. *Ann. Med. Hist.*, 1934, n.s. 6: 1–30, 143–149; 1935, n.s. 7: 428–437, 570–589; 1936, n.s. 8: 65–90; 1937, n.s. 9: 34–61; 1939, n.s. 11: 525–537.

Walshe, F. M. R., 1943. On the mode of representation of movements in the motor cortex, with special reference to "convulsions beginning unilaterally" (Jackson). *Brain*, 66: 104–139.

Walzer, R., 1929. *Galen on Jews and Christians.* Oxford University Press, Oxford.

Wasserstein, A. G., 1996. Death and the internal milieu: Claude Bernard and the origins of experimental medicine. *Perspect. Biol. Med.*, 39: 313–326.

Watson, J. B., 1924. *Behaviorism.* Norton, New York.

Watson, J. B., and Lashley, K. S., 1915. An historical and experimental study of homing in pigeons. *Publ. Carnegie Inst. Wash.*, 7: 7–60.

Wehrli, G. A., 1939. Trepanation in former centuries. *Ciba Symposium*, 1: 178–186.

Weiss, P. A., 1970. In: Schmitt, F. O., ed., *The Neurosciences: Second Study Program*. Rockefeller University Press, New York.

Weschler, L., 2006. *Everything that Rises: A Book of Convergences*. McSweeney's, San Francisco.

West, T. G., 1997. *In the Mind's Eye: Visual Thinkers, Gifted People with Dyslexia and Other Learning Difficulties, Computer Images, and the Ironies of Creativity*. Prometheus, Amherst, NY.

Whitaker, J. R., 1887. *Anatomy of the Brain and Spinal Cord*. E. and S. Livingstone, Edinburgh.

White, C., 1984. *Rembrandt*. Thames and Hudson, New York.

Whitehead, R. H., 1900. *The Anatomy of the Brain*. F. A. Davis, Philadelphia.

Whyte, L. L., 1975. Chirality. *Leonardo*, 8: 245–248.

Wiener, N., 1961 [1948]. *Cybernetics; or, Control and Communication in the Animal and the Machine*. MIT Press, Cambridge, MA.

Willis, T., 1664. *Cerebri anatome*. Martyn and Allestry, London.

Willis, T., 1683. *Two Discourses concerning the Soul of Brutes*. S. Pordage, trans. Dring, London.

Willis, T., 1684. *Dr. Willis's Practice of Physick*. S. Pordage, trans. Dring, London.

Wilkins, R. H., 1997. Neurosurgical Techniques: an Overview. In: Greenblatt, S. H., ed., *A History of Neurosurgery*. American Association of Neurological Surgeons, Park Ridge, IL.

Wilson, L. G., 1972. Galen. In: Gillespie, C. C., *Dictionary of Scientific Biography*. Scribner's, New York.

Winer, G. A., and Cottrell, J. E., 1996. Does anything leave the eye when we see? Extramission beliefs of children and adults. *Curr. Dir. Psychol. Sci.*, 5: 137–142.

Winer, G. A., and Cottrell, J. E., 2004. The odd belief that rays exit the eye during vision. In D. T. Levin, ed. *Thinking and seeing: Visual metacognition in adults and children*. MIT Press: Cambridge, MA.

Winer, G. A., Cottrell, J. E., Gregg, V., Fournier, J. S., and Bica, L. A., 2002. Fundamentally misunderstanding visual perception. Adults' belief in visual emissions. *Am. Psychol.*, 57: 417–424.

Winer, G. A., Cottrell, J. E., Karefilaki, K. D., and Chronister, M. C., 1996a. Conditions affecting beliefs about visual perception among children and adults. *J. Exp. Child Psychol.*, 61: 93–115.

Winer, G. A., Cottrell, J. E., Karefilaki, K. D., and Gregg, V. R., 1996b. Images, words, and questions: Variables that influence beliefs about vision in children and adults. *J. Exp. Child Psychol.*, 63: 499–525.

Winer, G. A., Rader, A. W., and Cottrell, J. E., 2003. Testing different interpretations for the mistaken belief that rays exit the eyes during vision. *J. Psychol.*, 137: 243–261.

Wiser, M., and Carey, S., 1983. When heat and temperature were one. In: Gentner, D., and Stevens, A. L., eds., *Mental Models*. Erlbaum, Hillsdale, NJ.

Wisner, E. P., 1965. Violations of Symmetry in Physics. *Sci. Am.*, December.

Wolff, U., and Lundberg, I., 2002. The prevalence of dyslexia among art students. *Dyslexia*, 8: 34–42.

Wölfflin, H., 1941. *Gedanken zur Kunstgeschichte*. Schwabe, Basel.

Woodall, J., 1639. *The Surgeons Mate, or Military and Domestique Surgery*. Bourne, London.

Woolam, D. H. M., 1958. Concepts of the brain and its functions in classical antiquity. In: Poynter, F. N. L., ed., *The History and Philosophy of Knowledge of the Brain and Its Functions*. Charles C. Thomas, Springfield, IL.

Woolsey, C. N., Settlage, P. H., Meyer, D. R., Spencer, W., Hamuy, T. P., and Travis, A. M., 1952. Pattern of localization in precentral and "supplementary" motor areas and their relation to the concept of a premotor area. In: *Association for Research in Nervous and Mental Disease*, vol. 30. Raven Press, New York.

Yamane, S., Kaji, S., and Kawano, K., 1988. What facial features activate face neurons in the inferotemporal cortex of the monkey? *Exp. Brain Res.*, 73: 1209–1214.

Yarbus, L., 1967. *Eye Movements and Vision*. Plenum, New York.

Young, A. W., and Ellis, H. D., 1976. An experimental investigation of developmental differences in ability to recognize faces presented to the left and right cerebral hemispheres. *Neuropsychology*, 14: 495–498.

Young, R. M., 1970. *Mind, Brain, and Adaptation in the Nineteenth Century*. Clarendon Press, Oxford.

Zago, S., Nurra, M., Scarlato, G., and Silani, V., 2000. Bartolomeo Panizza and the discovery of the brain's visual center. *Arch. Neurol.*, 57: 1642–1648.

Zeki, S. M., and Sanderman, D. R., 1976. Combined anatomical and electrophysiological studies on the boundary between the second and third visual areas of rhesus monkey cortex. *Proc. R. Soc. Lond. B*, 194: 555–562.

Zeller, E., 1955. *Outlines of the History of Greek Philosophy*. Meridian, New York.

Zola-Morgan, S., 1985. Localization of brain function: The legacy of Franz Joseph Gall (1758–1828). *Annu. Rev. Neurosci.*, 18: 359–383.

Index